Maren Daschner de Tercero

# Near-critical and supercritical hydrothermal flow synthesis of metal oxide nanoparticles and hybrid metal oxide nanoparticles presenting *clickable* anchors

# Near-critical and supercritical hydrothermal flow synthesis of metal oxide nanoparticles and hybrid metal oxide nanoparticles presenting *clickable* anchors

by
Maren Daschner de Tercero

KIT Scientific Publishing

Dissertation, Karlsruher Institut für Technologie (KIT)
Fakultät für Chemieingenieurwesen und Verfahrenstechnik, 2014
Tag der mündlichen Prüfung: 18. Februar 2014
Referent: Prof. Dr.-Ing. Michael Türk
Korreferent: Prof. Dr. rer. nat. Thomas Hirth

**Impressum**

KIT Scientific Publishing

Karlsruher Institut für Technologie (KIT)
KIT Scientific Publishing
Straße am Forum 2
D-76131 Karlsruhe

KIT Scientific Publishing is a registered trademark of Karlsruhe Institute of Technology. Reprint using the book cover is not allowed.

www.ksp.kit.edu



Print on Demand 2014

ISBN 978-3-7315-0262-3
DOI 10.5445/KSP/1000043063

# Near-critical and supercritical hydrothermal flow synthesis of metal oxide nanoparticles and hybrid metal oxide nanoparticles presenting *clickable* anchors

zur Erlangung des akademischen Grades eines
DOKTORS DER INGENIEURWISSENSCHAFTEN (Dr.-Ing.)

an der Fakultät für Chemieingenieurwesen und Verfahrenstechnik
des Karlsruher Instituts für Technologie (KIT)

genehmigte
DISSERTATION

von
Maren Daschner de Tercero, M.Sc.
aus Hamburg

| | |
|---|---|
| Referent: | Prof. Dr.-Ing. Michael Türk |
| Korreferent: | Prof. Dr. rer. nat. Thomas Hirth |
| Tag der mündlichen Prüfung: | 18. Februar 2014 |

*für die, die nicht gehört, übersehen oder missverstanden werden*

# Danksagung

Die vorliegende Arbeit entstand maßgeblich während meiner Tätigkeit als Wissenschaftliche Mitarbeiterin am Fraunhofer-Institut für Chemische Technologie (ICT) in der Abteilung Umwelt-Engineering (UE). Bei der Entwicklung der Idee(n) dieser Arbeit, ihrer Durchführung und Verfassung haben viele Personen mitgewirkt: direkt/indirekt, bewusst/unbewusst, angenehm/unangenehm, freiwillig/unfreiwillig. Sie alle waren wichtig. Wie diese Hände bzw. Köpfe alle ineinander gegriffen bzw. sich verwoben haben, ist jedoch nicht möglich hier in schwarz und weiß angemessen und richtig darzustellen. Hier dennoch ein Versuch in dem Wissen, dass sicherlich Pesonen vergessen werden, die nicht minder wichtig waren als andere, die hier das Glück oder Pech haben, auf diesen Seiten zu stehen.

Begonnen hat die Odyssee des Promotionsersuchens in Stuttgart bei Prof. Dr.-Ing. Hans Hasse am Lehrstuhl für Technische Thermodynamik und Thermische Verfahrenstechnik. Die Zeit dort hat mir die Schwierigkeiten der Handhabung von Proteinen – im speziellen Antikörper – und deren Aufreinigung verdeutlicht und den Blick geöffnet für die Verwendung von magnetischen Partikeln sowie Alternativen für Proteine in Therapeutik und Diagnostik. Der Wechsel in den „Großraum“ Karlsruhe hat mir vor allem das spannende Feld der überkritischen Fluide und der Hochdrucktechnik eröffnet und Anwendungen im Bereich Wasser ins Visier gerückt. Für die Möglichkeit, sich genauer mit den ersten beiden Feldern auseinanderzusetzen, möchte ich mich bei Prof. Dr. Thomas Hirth und Dr. Ulrich Fehrenbacher bedanken. Besonders schön ist, dass Herr Prof. Dr. Thomas Hirth, der den Anfang dieser Dissertation u.a. durch seine Begeisterung für das überkritische Wasser ermöglicht und geprägt hat, sich hier auch wieder am Ende bereit erklärt hat, sich durch die Übernahme des Korreferats zu engagieren. Vielen Dank! Bei der Festlegung der groben Richtung der Dissertation war die Unterstützung von Prof. Dr.-Ing. Ulrich Teipel nicht nur durch seine Expertise im Bereich „Partikeltechnologie“ besonders wertvoll. Ich bin ihm außerdem dankbar für sein beharrliches und frühes Drängen, die Ergebnisse in die Fachöffentlichkeit zu bringen. Dr. Stefan Jennewein, der insbesondere das hochspannende Thema der Glyconanopartikel eingebracht hat, welches ich sicherlich nicht mehr aus dem Auge verlieren werde, gilt ein spezieller Dank. Für die Möglichkeit, die Arbeiten zur Herstellung der *clickfähigen* eisenoxidischen Nanopartikel zu einem befriedigenden Entwicklungspunkt bringen zu können, möchte ich mich bei der Leitung von UE, Rainer Schweppe, bedanken.

Ein großer Dank gilt Prof. Dr.-Ing. Michael Türk für sein Interesse und Engagement an dem Thema und seine angenehme und unkomplizierte Art, mich als externe Doktorandin am Institut für Technische Thermodynamik und Kältetechnik des Karlsruher Instituts für Technologie (KIT) zu betreuen.

Ganz besonders möchte ich mich bei Dr. Leonie Barner bedanken, die an ganz unterschiedlichen Stellen, in verschiedenen Situationen und auf vielfältige Weise wichtige Impulse zum Voranbringen und Fertigstellen der Arbeit gesetzt hat.

Basis der vorliegenden Arbeit war eine funktionierende Hochdruckanlage. Für die Unterstützung bei der Planung, dem Bau und den vielen Umbauten (und der Leckortung) möchte ich mich bei den Hochdruckspezialisten von UE – Jürgen Graf, Stefan Unser und Detlef Rückert – bedanken, sowie Lothar Heck und den Umweltsimulanten für ihre Unterstützung beim Aufbau der Mess- und Regeltechnik und Günther Mehring für die angenehme und dabei präzise Druckkalibrierung der entscheidenden Drucksensoren.

Ebenso wichtig wie die Herstellung der Nanopartikel war deren Charakterisierung. Im Hinblick auf die so wichtigen Partikelgrößenverteilungsbestimmungen der unzähligen Dispersionen möchte ich mich insbesondere bei Irma Mikonsaari für die Idee und die Möglichkeit der Nutzung der Scheibenzentrifuge (DCS) am ICT bedanken. Dank Prof. Dr. Gerd Irmer und Christian Röder vom Institut für Theoretische Physik an der TU Bergakademie Freiberg und deren Ramanspektroskop, sowie Dr. Michael Herrmann, Dr. Harald Fietzek und Ulrich Förter-Barth (Durchführung und Auswertung der PXRD-Messungen), am ICT konnten wir Licht in das Dunkel der strukturellen Möglichkeiten der synthetisierten Eisenoxide bringen. Dank Dr. Michael Bruns und den Röntgenelektronenspektroskopen (XPS) vom Institut für Angewandte Materialien (aktuell IAM-WPT, über die Namensänderungen habe ich im Laufe der Zeit den Überblick verloren), Dr. Alexandra Greiner und dem konfokalem Laserfluoreszenzmikroskop (LSCM) vom Zoologischen Institut, Zell- und Neurobiologie, sowie Dr. Christian Kübel und dem Transmissionselektronenmikroskop (TEM) vom Institut für Nanotechnologie (INT) – alle KIT, konnten wir die Oberflächenchemie, -funktionalisierung und Morphologie der Partikel unter die „Lupe" nehmen. Hervorheben möchte ich Dr. Michael Bruns, der trotz der vielen Unterbrechungen immer bereit und aufgeschlossen war, spannenden und weniger spannenden Phänomenen auf den Grund zu gehen.

Außerdem möchte ich mich bei folgenden Personen bedanken, die mich durch die Durchführung zahlreicher weiterer Analysen und Labortätigkeiten unterstützt haben: Dr. Jördis Eisenblätter (NMR und Labor), Gabriela Gromer (ICP-OES), Katrin Hennig (TOC), Christiane Lang (Synthese von Referenzsubstanzen), Sonja Lauinger (ICP-OES), Desiree Westermann (Click-Synthesen uvm.), Dr. Detlef Schmiedel (LC-MS), Heike Schuppler (TG), Wenka Schweikert (IR und Raman), Winfried Send (TEM, Institut für Elektronenmikroskopie, KIT), Gabriela Wolf (IC), Dr. Beatrice Tübke (Leitung Analytik UE) und Virginie Zöller (LC-MS).

Bei Dr. Anja Goldmann und Dr. Andreas Kaiser vom Institut für Technische Chemie und Polymerchemie, KIT, möchte ich mich für Diskussionen und Impulse zur Click-Chemie bzw. zu funktionalisierten magnetischen Partikeln bedanken, sowie Dr. Nicolas Zydziak zum Thema „$TiO_2$“. Ebenso herzlichen Dank an Prof. Dr. Matthias Franzreb und Jonas Wohlgemuth vom Institut für Funktionelle Grenzflächen (IFG), KIT, die mir ermöglicht haben, orientierende Untersuchungen zu den magnetischen Eigenschaften der Partikel mittels eines „alternating gradient magnetometer“ (AGM) durchzuführen.

Ute Pohsner, die mich persönlich ausgesprochen wertvoll in meinen Jahren am ICT unterstützt hat, sowie Otto Grosshardt, Gerd Unkelbach, Birgit Gartner und Ute Laubach, die mich in UE positiv begleitet haben, möchte ich an dieser Stelle danken.

Ohne die Unterstützung studentischer Mitarbeiter und Mitarbeiterinnen, die die sehr unterschiedlichen Phasen der Arbeit mitgestaltet haben und erdulden konnten, wäre diese Arbeit nicht enstanden. Für ihr Engagement, sowie ihr Vertrauen und die Bereitschaft sich auf mich und die Hochdrucktechnik einzulassen, möchte ich mich bei Alexander Dresel, Johanna Schubert, Franziska Keilmann, Claudia Mönius, Isaí Gonzáles Martínez, Christian Roßmann und Marcus Zang, der auch als späterer Kollege immer wieder bereit war im Notfall einzuspringen, bedanken.

Der Fraunhofer–Gesellschaft und der Karlsruher Nano Micro Facility (KNMF), einer Helmholtz Infrastruktureinrichtung am KIT, danke ich für die finanzielle Unterstützung.

Ein großer Dank geht außerdem an Dr. Matthias Kautt und Dr. Verena Höckele vom Projektträger Karlsruhe, die bereit waren, mich auch als „Unfertige“ aufzunehmen.

Hierbei möchte ich mich auch bei allen meinen neuen Kollegen vom PTKA-WTE bedanken, die mir in vielen Wochen ein so schönes und angenehmes Gegengewicht zum Schreiben von Manuskripten und der Diss zu Hause bereitet haben.

Herzlichen Dank geht an Dr. Leonie Barner, Christian Röder, Christel und Rainer Sturm und Luis, meinem Mann, für die kritische und konstruktive Durchsicht der schriftlichen Fassung(en) der Arbeit.

Besonders möchte ich Dunja K., Michael S., Wiebke G., Frank L., Katrin D., den MSGs und SSTs sowie LATE danken, die für mich auf sehr unterschiedliche Weise in der Zeit sehr wichtig waren. Meine Nia-Trainerin, -Kollegin und -Teilnehmer halfen mir Präzision und Kreativität in einer ganz anderen Form zu nutzen und zu fördern.

Zu guter Letzt: keiner hat so direkt die Phase des „Dissens“ von der ersten Stunde an begleitet wie mein Sohn. Von und mit ihm habe ich persönlich in all den Jahren am meisten gelernt. Ihm gilt mein ganz besonderer und tiefer Dank.

Karlsruhe, November 2013

Maren Daschner de Tercero

# Contents

# Zusammenfassung

Nah- und überkritisches Wasser stellt ein attraktives Medium zur Synthese von metalloxidischen Nanopartikeln dar. Der Prozess der kontinuierlichen hydrothermalen Synthese basiert auf einem einfachen Prinzip: dem Vermischen eines kalten wässrigen Metallsalzstromes mit heißem komprimiertem Wasser (siehe Abbildung 1.1; Adschiri et al., 1992). Hierbei wird ausgenutzt, dass sich die Eigenschaften des Wassers – insbesondere die Permittivität – zwischen Raumbedingungen und Bedingungen in der Nähe des kritischen Punktes ($p$ = 22,1 MPa, $T$ = 647,1 K) stark unterscheiden. Im nah- und überkritischen Bereich hat Wasser ungewöhnliche Eigenschaften. Makroskopisch verhält es sich wie ein unpolares Lösungsmittel, die einzelnen Moleküle sind jedoch weiterhin polar (Kruse and Dinjus, 2007). Die speziellen und einstellbaren Eigenschaften des Wassers um den kritischen Punkt erlauben des Weiteren die Herstellung von in situ funktionalisierten metalloxidischen Nanopartikeln. Dies ermöglicht eine gezieltere Gestaltung des Partikelbildungsprozesses in Hinblick auf die Größe, Morphologie und Oberflächenchemie der Partikel im Vergleich zur kontinuierlichen hydrothermalen Synthese von nicht funktionalisierten Partikeln (z.B. Byrappa et al., 2008; Mousavand et al., 2006; Taguchi et al., 2012; Ziegler et al., 2001).

Im Rahmen dieser Dissertation wurde sowohl die kontinuierliche hydrothermale Synthese von funktionalisierten als auch nicht funktionalisierten Nanopartikeldispersionen vorangetrieben. Hierbei wurden folgende Ziele verfolgt:

1. Steigerung des Prozessverständnisses hinsichtlich der Einflüsse der Prozessvariablen auf die Eigenschaften der Produktdispersionen.
2. Erweiterung der Anwendungsmöglichkeiten der kontinuierlichen hydrothermalen Partikelsynthese durch eine in situ Funktionalisierung der metalloxidischen Partikel mit *clickfähigen** Molekülen.
3. Aufzeigen eines einfachen Syntheseweges zur Bereitstellung von Glyconanopartikeln†.

*(Moleküle, die) *Click*-Reaktionen eingehen können, welche als einfach, effizient und vielfältig anwendbar bei modularen Synthesestrategien definiert sind (Kolb et al., 2001).

†Nanopartikel, die an ihrer Oberfläche Oligosaccharide tragen und dadurch die sog. "Glycocalyx", welche die Oberfläche zahlreicher Zelltypen bedeckt, nachahmen (de la Fuente et al., 2001).

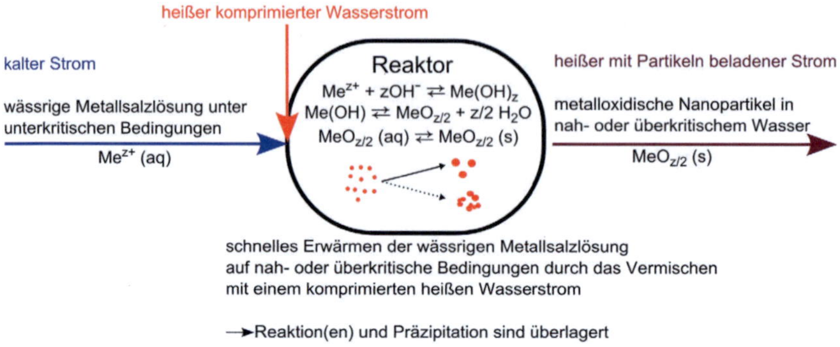


**Abbildung 0.1:** Schematische Darstellung der kontinuierlichen hydrothermalen Synthese von metalloxidischen Nanopartikeln.

## Konzeption, Bau und Betrieb der Pilotanlage

Die Grundlage für die Verfolgung der Ziele der vorliegenden Dissertation bildete die Konzeption, der Bau, der Betrieb und die Optimierung einer Pilotanlage zur kontinuierlichen hydrothermalen Herstellung von metalloxidischen Nanopartikeln unter nah- und überkritischen Bedingungen. Es wurde eine Pilotanlage mit einem funktionierendem Mess-, Regel- und Datenerfassungssystem unter Berücksichtigung der speziellen und notwendigen Vorsichts- und Sicherheitsmaßnahmen, die zum Betreiben von Anlagen unter hohem Druck und hoher Temperatur notwendig sind, gebaut ($T_{max} = 773$ K and $p_{max} = 38$ MPa). In der Detailplanung der Hochdruckanlage wurde ein besonderes Augenmerk auf folgende Aspekte gerichtet:

- Die Realisierung von hohen Gesamtmassenströmen ($\dot{m}_{max} = 12\ \mathrm{kg\,h^{-1}}$) im Vergleich zu den aus der Literatur bekannten Pilotanlagen. Hierdurch konnten sehr kurze Verweilzeiten erzielt werden. Des Weiteren ermöglichte dies Untersuchungen zum Einfluss der Massendurchflüsse auf den Partikelbildungsprozess über einen großen Bereich an mittleren Strömungsgeschwindigkeiten.
- Eine komfortable Umstellung zwischen verschiedenen Mischer- und Reaktorgeometrien. Dies ermöglichte das Vermischen von (zusätzlichen) Strömen an verschiedenen Positionen und unter Verwendung unterschiedlicher Geometrien innerhalb der Reaktionszone.
- Ein pulsationsarmer Betrieb der Pilotanlage aufgrund der starken Änderungen der Reaktionsbedingungen bereits bei kleinen Druckschwankungen.

# Herstellung von ceroxidischen Nanopartikeldispersionen

Ceroxid wurde als Modellsystem zum Einfahren und Testen der neu aufgebauten Pilotanlage ausgewählt. Die ersten orientierenden Versuche dienten (a) dem Testen der Reproduzierbarkeit und Zuverlässigkeit des Prozesses unter gleichen Einstellbedingungen und (b) dem Testen der Zuverlässigkeit und Stabilität des Prozesses bei unterschiedlich hohen Cer-Gehalten. Wässrige Cernitratlösungen dienten hierbei als Ausgangslösung.

Diese Versuche lieferten elektrostatisch stabilisierte Ceroxid-Nanopartikeldispersionen bei Umsätzen $> 99.8\%$ und mittleren Verweilzeiten $< 1$ s. Die mittlere Kristallitgröße der ceroxidischen Nanopartikel, gemessen mittels Röntgenpulverdiffraktometrie, betrug 30–45 nm. Mit Hilfe der differentiellen zentrifugalen Sedimentationsanalyse, einer Messmethode, die zugleich leistungsfähig für die Messung von Partikel- bzw. Agglomeratgrößenverteilungen als auch praktisch zu bedienen ist, konnte gezeigt werden, dass die ceroxidischen Nanopartikel bei niedrigen Cer-Gehalten vorwiegend als Einzelpartikel in Dispersion vorlagen, während sich bei steigenden Cer-Gehalten Aggregate bzw. Agglomerate bildeten. Dabei ist zu beachten, dass die mittlere Kristallitgröße der Nanopartikel keine signifikante Abhängigkeit vom Cergehalt, der über drei Größenordnungen variiert wurde, aufwies.

Eine verhältnismäßig geringe Reproduzierbarkeit der Versuchsergebnisse hinsichtlich der mittleren Kristallitgröße und gelegentliches Zusetzen der Anlage aufgrund von Ablagerungen am Mischpunkt waren Hinweise für die Notwendigkeit einer Optimierung des Mischers. Dies wurde bei den Versuchen zur Herstellung eisenoxidischer Nanopartikel berücksichtigt.

# Herstellung von eisenoxidischen Nanopartikeldispersionen

Elektrostatisch stabilisierte eisenoxidische Nanopartikeldispersionen wurden aus wässrigen Eisen(II)acetat, Eisen(III)nitrat und Mischungen dieser zwei Eisensalze mit oder ohne Zugabe von den entsprechenden Säuren ($CH_3COOH$ bzw. $HNO_3$) hergestellt. Die Charakterisierung der hergestellten Dispersionen erfolgte mittels differentieller zentrifugaler Sedimentationsanalyse, elektrophoretischer Lichtstreuung, Transmissionselektronenmikroskopie (TEM), Röntgenpulverdiffraktometrie und Ramanspektroskopie. Für letztere wurde ein modifizierter Analysenaufbau zur einfachen Untersuchung konzentrierter eisenoxidischer Dispersionen verwendet. Dispersionen, die aus reinem Eisennitrat hergestellt wurden, wiesen eine bimodale Partikelgrößenverteilung sowohl in Hinblick auf die Primärpartikel als auch Agglomerate auf (vgl. Abbildung 0.2). Allerdings konnte durch Erhöhung des Gesamtmassenstromes sowie des Anteils des Metallsalzstromes an diesem der Anteil der Fraktion der großen Partikel und Agglomerate (im Bereich von 30 nm bzw. 65 nm) zu Gunsten der kleinen Partikel und Agglomerate (im Bereich von 7 nm bzw. 20 nm) fast vollständig reduziert werden.

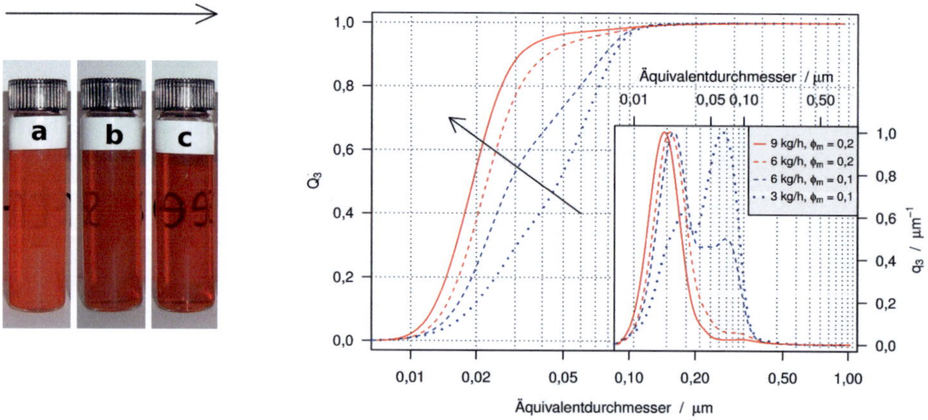


**Abbildung 0.2:** Charakterisierung der wässrigen Eisenoxiddispersionen, hergestellt bei verschiedenen Gesamtmassenströmen ($\dot{m}_{\text{gesamt}}$) und Mischungsverhältnissen ($\phi_m$, Anteil kalter Metallsalzstrom im Gesamtmassenstrom) unter Verwendung von Eisennitrat mit einer Konzentration an Eisen im Produkt ($c_{\text{Produkt Fe}}$) von 0,0055 mol/L. (Links) Erscheinungsbild von Dispersionen, synthetisiert bei einem $\dot{m}_{\text{gesamt}}$ von (a) 3,04 kg/h, (b) 6,07 kg/h bzw. (c) 9,11 kg/h und bei einem konstanten $\phi_m$ von 0,2. (Rechts) Integrale und differentielle Partikelgrößenverteilung ($Q_3$ and $q_3$) aus differentiellen Sedimentationsanalysen unter Verwendung einer Scheibenzentrifuge. Die Pfeile zeigen in Richtung steigendem $\dot{m}_{\text{gesamt}}$ und/oder $\phi_m$.

Ramanspektroskopische Untersuchungen zeigten, dass die Nanopartikel aus einer Mischung aus $\alpha$-$Fe_2O_3$ und $\gamma$-$Fe_2O_3$ und/oder Ferrihydrit bestanden. Des Weiteren konnte durch Kombination der Analysen in Dispersion mit Analysen getrockneter Nanopartikel gezeigt werden, dass sowohl die Struktur als auch die Größe der Nanopartikel sich bei der Trocknung ändern können. Das bedeutet, dass falsche Schlussfolgerungen gezogen werden können, wenn die eisenoxidischen Partikel nicht direkt in der Dispersion untersucht werden. Weitere ramanspektroskopische Untersuchungen an eisenoxidischen Nanopartikeln in Dispersion sind jedoch notwendig, um die Ramanspektroskopie hin zu einer aussagekräftigeren Methode bei der Bestimmung von Struktur und Größe eisenoxidischer Nanopartikeln zu entwickeln. Erkenntnisse in diesem Kontext sind wegen der allgemeinen Bedeutung von Eisenoxiden nicht nur für die hydrothermale Nanopartikelsynthese wichtig.

Signifikant engere Partikel- bzw. Agglomeratgrößenverteilungen mit Primärpartikelgrößen $< 10$ nm und einheitlicher Struktur konnten aus Mischungen von Eisen(II)acetat und Eisen(III)nitrat anstelle von reinem Eisen(III)nitrat unter ansonsten gleichen Strömungsverhältnissen erzielt werden (siehe Abbildung 0.3). Des Weiteren zeigten Versuche unter systematischer Veränderung der Zusammensetzung der Ausgangslösung das Potential zur kontinuierlichen Variation der Partikel- und Agglomeratgrößenverteilungen der Produktdispersionen. Auch wenn weitere Studien notwendig sind, um zwischen den verschiedenen Effekten zu unterscheiden, die zu einer Verschiebung der gemessenen Partikelgrößenverteilungen führen, erscheint es zweckmäßig, die Eduktzusammensetzung als weitere Variable bei der Suche nach optimalen Prozessbedingungen einzubeziehen. Dies ist umso wichtiger, da die Eduktzusammensetzung auch einen Effekt auf die Sensitivität der Eigenschaften der Produktsuspensionen in Bezug auf Schwankungen in den Prozessbedingungen hat, was ein wichtiger Aspekt beim Scale-up ist.

Zu beachten ist, dass im Rahmen der Versuche zur Herstellung von eisenoxidischen Nanopartikeln ein anderer Mischer (T-Stück mit einem konstanten Innendurchmesser) als für die Herstellung von ceroxidischen Nanopartikeln (Standard Hochdruck T-Stück mit abrupten Änderungen im Innendurchmesser) verwendet wurde. Selteneres Zusetzen der Anlage durch Ablagerungen im Mischer waren das Ergebnis dieser Veränderung der Mischergeometrie, die, zusammen mit optimierten Strömungsverhältnissen, einen verlässlichen Langzeitbetrieb der Anlage ermöglichten. Die Nutzung eines hochsensiblen Materialsystems wie die Eisenoxide/ -oxohydroxide in Kombination mit dem Einsatz von Methoden zur Analyse der

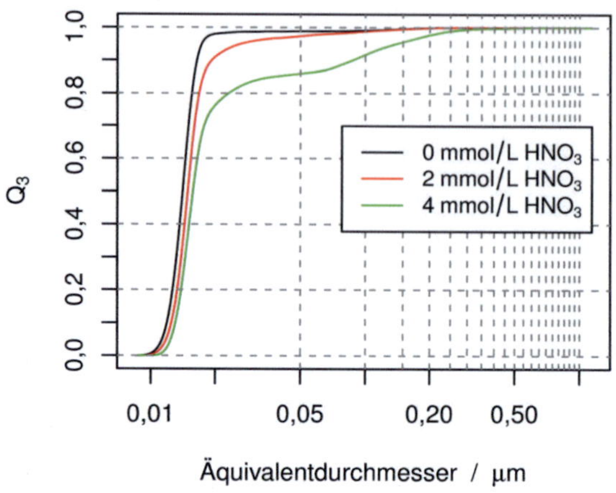


**Abbildung 0.3:** Integrale Partikelgrößenverteilungen ($Q_3$) von Eisenoxiddispersionen hergestellt unter Verwendung von Mischungen aus Eisen(II)acetat (65%) und Eisen(III)nitrat (35%) unter der Zugabe von 0,003 mol/L Essigsäure und verschiedenen Mengen an Salpetersäure mit $c_{\text{Produkt Fe}}$ = 0,001 mol/L, $\phi_m$ = $0,2$ und $\dot{m}_{Gesamt} = 6,07$ kg/h.

Partikel-/Agglomeratgrößenverteilungen direkt in Suspension bieten einen guten Ausgangspunkt bei der Optimierung der kontinuierlichen hydrothermalen Partikelsynthese, wie z. B. bei der Suche nach optimalen Mischergeometrien.

## In situ Funktionalisierung von eisenoxidischen Nanopartikeln

Neben der Herstellung von elektrostatisch stabilisierten Dispersionen konnten eisenoxidische Nanopartikeldispersionen mit unterschiedlichen Eigenschaften (u.a. in Hinblick auf die Farbe, Kristallstruktur und Partikelgrößenverteilung) im kontinuierlichen Hochdruckprozess unter Zugaben von organischen Molekülen erzielt werden. Hierbei wurde sowohl die Anordnung der Mischer, die Temperatur in der Reaktionszone, das Ausgangsmaterial (Eisensalz) sowie die Art der verwendeten organischen Moleküle variiert. Ziel war hierbei, in situ funktionalisierte Nanopartikel zu synthetisieren, die geeignet sind für eine Sekundärfunktionalisierung mittels

*Click*-Chemie. Carbonsäuren mit einer endständigen Alkingruppe wurden zur Oberflächenmodifikation eingesetzt, da hiermit die Möglichkeit zur einfachen Kopplung mit einem Azid-funktionalisiertem Molekül mittels der am meisten verwendeten und bekanntesten *Click*-Reaktion, der Kupfer(I)-katalysierten Azid-Alkin-Cycloaddition (CuAAC), gegeben ist (Such et al., 2012). Des Weiteren wurde die Stabilität der *click-fähigen* Liganden unter den hydrothermalen Bedingungen untersucht. Hierbei zeigte sich, dass die Alkingruppe ausreichend stabil ist. Weitere Untersuchungen sind jedoch notwendig, um die Zersetzungswege aufzuklären und um Möglichkeiten zur weiteren Stabilisierung der Alkinsäuren bei der in situ Funktionalisierung zu eruieren.

Partikel, die in Anwesenheit von Carbonsäuren synthetisiert wurden, konnten zumindest teilweise in eine Toluolphase transferiert werden, was als Indikator für das Eingehen einer Physi- oder Chemisorption der organischen Moleküle mit der Oberfläche der eisenoxidischen Nanopartikel anzusehen ist. Das Ausmaß des Phasentransfers von der wässrigen in die organische Phase korrelierte mit der Länge der Carbonsäure und der Art der endständigen funktionellen Gruppe.

Partikel, die in Anwesenheit von 10-Undecinsäure synthetisiert wurden, bildeten nach Abtrennung von größeren Partikeln bzw. Agglomeraten mittels Zentrifugation eine stabile Dispersion in Ammoniakwasser (siehe Abbildung 0.4a). Dies lässt eine Kombination von sterischen und elektrostatischen Mechanismen vermuten, die mit der Bildung einer Doppel- oder Mehrfachschicht aus 10-Undecinsäuremolekülen und endständigen Carboxylat-Ionen einherginge. TEM-Aufnahmen dieser Dispersion zeigen Primärpartikel in Größen von 4-7 nm als Bestandteile von Agglomeraten/Aggregaten (siehe Abbildung 0.4b). Die Charakterisierung der Partikelgrößenverteilungen mittels differentieller zentrifugaler Sedimentationsanalyse zeigte, dass die Primärpartikel in Form von Aggregaten/Agglomeraten mit einem mittleren Äquivalentdurchmesser von 21,5 nm (massenbezogener Medianwert) vorlagen, wobei 80% der Partikel zwischen 13,5 und 30,5 nm detektiert wurden. Eine kleine Fraktion an Partikeln mit Durchmesseren um die 250 nm wurde außerdem gemessen (Abbildung 0.4c).

Um die Anwesenheit von *clickfähigen* funktionellen Gruppen an der Oberfläche der Partikel zu verifizieren, wurde die Kupfer(I)-katalysierte Azid-Alkin-Cycloaddition (CuAAC) mit den Alkin-funktionalisierten Partikeln und einem Azid-funktionalisierten Fluoreszenzfarbstoff (ein Derivat von Rhodamin-B) durchgeführt. Mit Hexansäure funktionalisierte Partikel (keine Alkyl-Gruppen) und Partikel, die

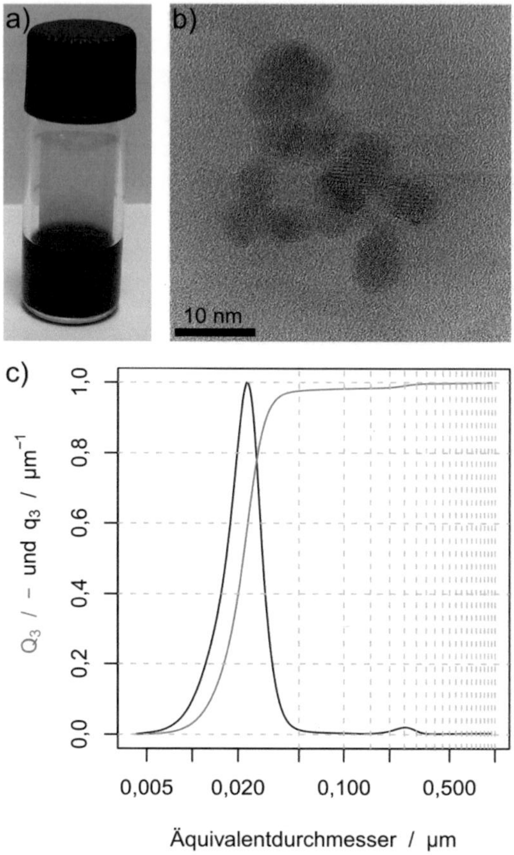


**Abbildung 0.4:** Charakterisierung einer wässrigen Dispersion von mit 10-Undecinsäure funktionalisierten eisenoxidischen Nanopartikeln. (a) Erscheinungsbild der Dispersion nach mehreren Monaten; (b) TEM-Aufnahme von Partikeln der Dispersion; (c) integrale und differentielle Partikelgrößenverteilung ($Q_3$ and $q_3$) aus differentiellen Sedimentationsanalysen unter Verwendung einer Scheibenzentrifuge.

der CuAAC nicht unterzogen wurden, dienten als (negative) Kontrollproben. Aufkonzentrierte Dispersionen der mehrfach gewaschenen *geclickten* Partikel wurden mit Hilfe eines konfokalen Laser-Scanning-Mikroskops (LSCM) untersucht, wobei ein starkes Fluosreszenzsignal festgestellt wurde. Bei den (negativen) Kontrollproben konnte hingegen kein Fluoreszenzsignal detektiert werden (siehe Abbildung 0.5a). Partikel, die mit einer Mischung aus *clickfähigen* und nicht-*clickfähigen* Molekülen funktionalisiert waren sowie größere Partikel mit einem kleineren Oberfläche-zu-Volumen-Verhältnis, zeigten deutliche, jedoch geminderte Fluoreszenzsignale. Um die *Clickfähigkeit* der Partikel weiter zu untermauern, wurde die Bildung des 1,2,3-Triazolrings als Folge der CuAAC zwischen den funktionalisierten Partikeln und der Modellsubstanz Benzylazid mittels Röntgenphotoelektronenspektroskopie (XPS) untersucht. Hierbei wurde ein charakteristisches Multiplett mit mindestens zwei Peaks (siehe Abbildung 0.5b), die zu chemisch verschiedenen N-Komponenten ($\underline{N}=\underline{N}-N$ und $N=N-\underline{N}$) zuzuordnen sind, beobachtet, was ein Nachweis für die Bildung des 1,2,3-Triazolrings ist (Ciampi et al., 2007, 2011; James et al., 2011).

Neben der prinzipiellen Demonstration der Herstellung von clickfähigen metalloxidischen Nanopartikeln mittels in situ Funktionalisierung im kontinuierlichen hydrothermalen Prozess wurde die Synthese von Glyconanopartikeln auf Basis der primärfunktionalisierten eisenoxidischen Partikel verfolgt. Glyconanopartikel stellen eine Art von Hybridnanopartikeln dar, die aufgrund des hohen, aber lang unterschätzten Potenzials von Oligosacchariden von zunehmender Bedeutung sind. Diese können eine Vielzahl von biologischen Funktionen entfalten, wenn sie geclustert an einer Oberfläche orientiert vorliegen. Beispiele für potentielle Anwendungen von Glyconanopartikeln sind Krebsimpfstoffe (Ojeda et al., 2007), antivirale Wirkstoffe (Martínez-Ávila et al., 2009) und In-vivo-Diagnostika (van Kasteren et al., 2009). Im Rahmen dieser Arbeiten konnte die Herstellung eines Glyconanopartikelmodells durch *Click*-Reaktion zwischen den Alkin-funktionalisierten eisenoxidischen Partikeln und Monosacchariden mit Azid-Linkern gezeigt werden.

---

**Abbildung 0.5: (nächste Seite)** Untersuchung der *Clickfähigkeit* der in situ funktionalisierten Partikel mittels (a) LSCM unter Verwendung eines Azid-funktionalisierten Fluoreszenzfarbstoffes (ein Dericat von Rhodamin B) und (b) XPS unter Verwendung einer Modellsubstanz (Benzylazid). Die Verbindungslinien zwischen den organischen Molekülen und der Oberfläche der Partikel dienen der Veranschaulichung. Die tatsächliche Bindungskonfiguration ist unbekannt. Das Fluoreszenzbild der Agglomerate in der Waschlösung wurde bei gleichen Leistungseinstellungen wie die anderen Fluoreszenzbilder aufgenommen. Der Kontrast wurde zur Verdeutlichung nachträglich erhöht.

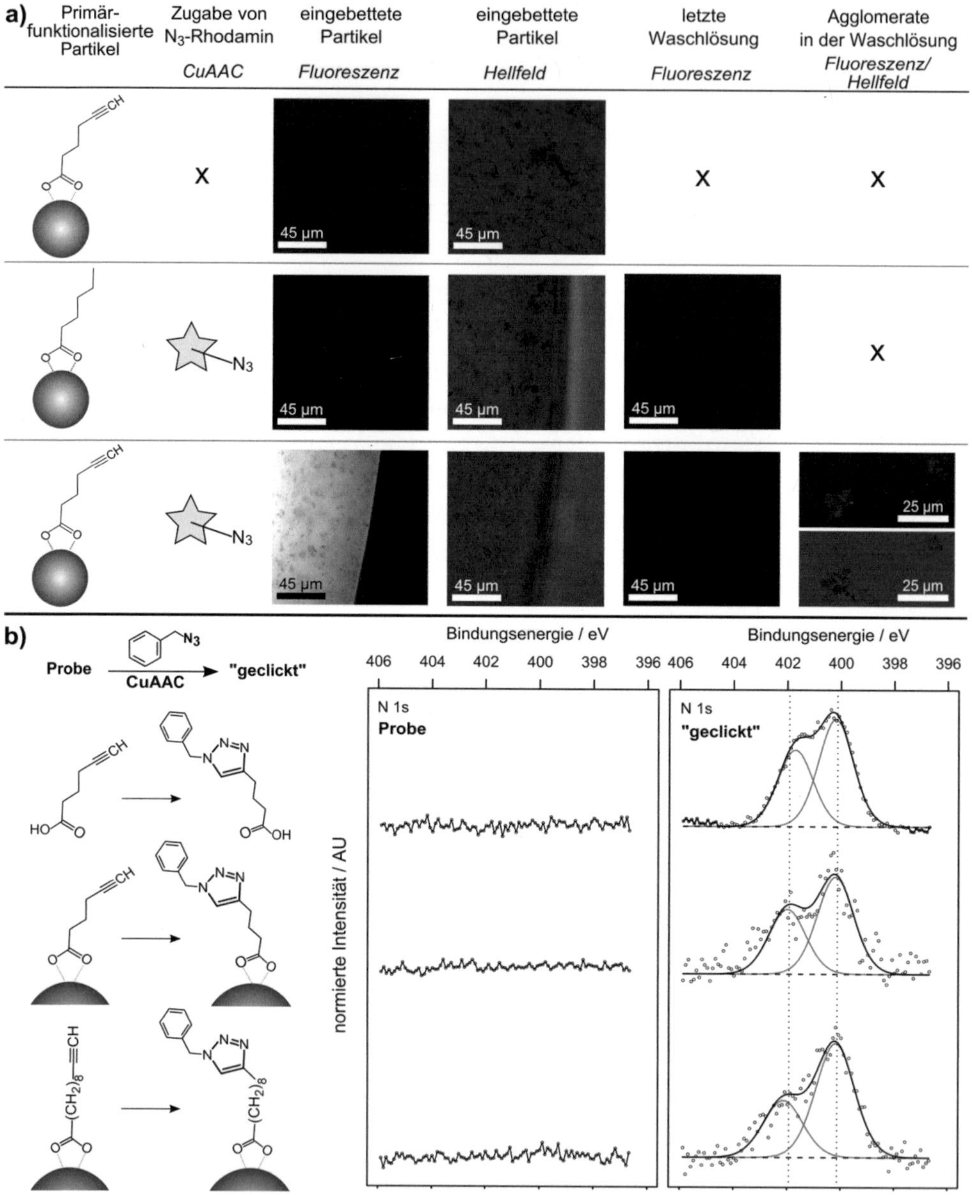

a)
Primär-funktionalisierte Partikel
Zugabe von N3-Rhodamin
CuAAC
eingebettete Partikel
Fluoreszenz
eingebettete Partikel
Hellfeld
letzte Waschlösung
Fluoreszenz
Agglomerate in der Waschlösung
Fluoreszenz/ Hellfeld
X
45 µm
25 µm
N3
b)
Probe
CuAAC
"geclickt"
Bindungsenergie / eV
406
404
402
400
398
396
N 1s
Probe
N 1s
"geclickt"
normierte Intensität / AU

Betrachtet man die Ergebnisse zur Herstellung von in situ funktionalisierten clickfähigen eisenoxidischen Nanopartikeln zusammen mit der Breite und Vielfältigkeit an metalloxidischen Nanopartikeln, die mittels eines kontinuierlichen hydrothermalen Prozesses erzielt werden können und den zahlreichen Arten an Molekülen, die über *Click*-Reaktionen zugänglich sind, birgt dieser Ansatz das Potenzial zur Herstellung von einer Vielzahl an Kern-Schale-Nanopartikeln mit bekannten, anvisierten und noch zu entdeckenden Eigenschaften (Adzima and Bowman, 2012; Hayashi and Hakuta, 2010). Die Verwendung von nanostrukturierten organisch-anorganischen Hybridmaterialien ist in zahlreichen Technologien in den Bereichen Wasser, Energie, Umwelt, Gesundheit, Kommunikation und Mobilität von (zunehmender) Bedeutung. Einige Verwendungsmöglichkeiten von Kern-Schale-Nanopartikeln sind in Abbildung 0.6 gezeigt, in der des Weiteren die maßgeblichen Beiträge dieser Dissertation herausgestellt sind.

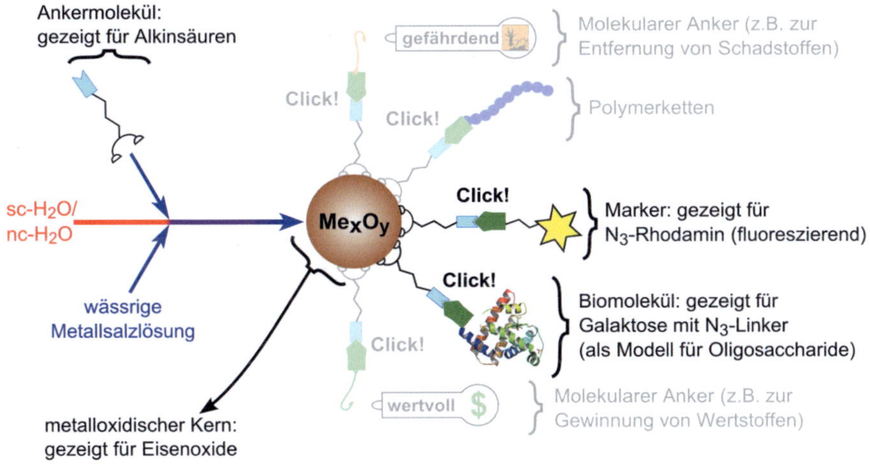


**Abbildung 0.6:** Schematische Darstellung des Prozesses zur Herstellung von metalloxidischen Nanopartikeln mit *clickfähigen* Ankermolekülen und ausgewählte potentielle Verwendungsmöglichkeiten. Hervorgehoben sind die Aspekte, die im Rahmen der vorliegenden Dissertation behandelt wurden.

# Chapter 1

# Introduction

Supercritical water has very unusual and tunable properties. Macroscopically it behaves like a non-polar solvent but the single molecules are still polar (Kruse and Dinjus, 2007). Properties related to mass transport are similar to those of gases, whereas the density and solvating capability are more similar to liquids (Lam et al., 2008). Together, these properties make near-critical and supercritical water an attractive medium for the synthesis of metal oxide nanoparticles. The continuous hydrothermal synthesis process is based on the rapid mixing of a cold aqueous metal salt solution with hot compressed water, as shown schematically in Figure 1.1, and exploits the strong differences of the properties of water—in particular its permittivity—between ambient conditions and conditions around the critical point ($p_c$ = 22.1 MPa, $T_c$ = 647.1 K; Adschiri et al., 1992).

The continuous hydrothermal synthesis of metal oxide nanoparticles in near-critical and supercritical water has been investigated for more than 20 years (Adschiri et al., 1992). Its versatility has been shown through the preparation of nanoparticles from a variety of material systems ranging from simple metal oxides such as $CeO_2$ (Hakuta et al., 2004; Kellici et al., 2010), $\gamma$-$Al_2O_3$ (Noguchi et al., 2008), $ZrO_2$ (Aimable et al., 2009; Hakuta et al., 2004), and iron oxides (Lam et al., 2008; Sue et al., 2010) to complex mixed metal oxides such as $Ca_{0.8}Sr_{0.2}Ti_{1-x}Fe_xO_{3-\delta}$ ($x = 0.1 - 0.3$) (Lu et al., 2008), $ZnGa_2O_4$:$Mn^{2+}$ (Hayashi et al., 2013), and $(Y_{2.7}Tb_{0.3})Al_5O_{12}$ (Hakuta et al., 2003). This process allows the rapid production of highly crystalline particles in an environmentally-benign fluid while tuning the particle size and morphology just by varying temperature, pressure and residence time (Adschiri et al., 1992; Bremholm et al., 2009a). Continuous hydrothermal synthesis is also suitable for the large-scale production of functional metal oxide nanoparticles, based on the recent development of highly-efficient mixing units for the nc-$H_2O$ and sc-$H_2O$ reaction and solvent environments (e.g. Gruar et al., 2013; Hong et al., 2013; Kawasaki et al., 2010; Lester

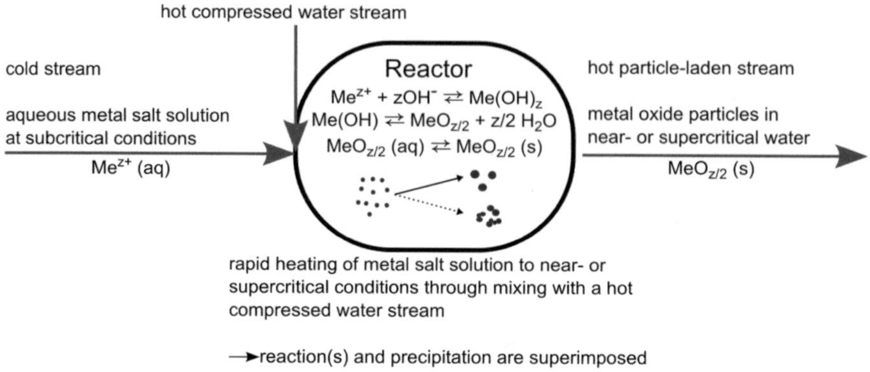


**Figure 1.1:** Simplified scheme of the continuous hydrothermal metal oxide nanoparticle synthesis.

et al., 2006; Sue et al., 2010), time-resolved in situ studies using synchrotron X-ray scattering to gain insight into the particle formation process (Bremholm et al., 2009b, 2008; Middelkoop et al., 2009), and automated high-throughput processes for the rapid screening of phase diagrams (Kellici et al., 2010; Lin et al., 2010; Weng et al., 2009). Furthermore, the continuous supercritical hydrothermal synthesis of metal oxide nanoparticles is especially attractive when the particles are meant to be used as aqueous suspensions because the addition of organic solvents, strong bases or surfactants is generally not necessary to obtain highly crystalline metal oxide nanocrystals (Lam et al., 2008; Sue et al., 2011).

This process can also be modified to produce organic-inorganic hybrid particles by in situ functionalization with organic molecules such as thiols, alcohols, aldehydes, carboxylic acids or amines (Mousavand et al., 2007; Ziegler et al., 2001) as organic modifiers in homogeneous nc-$H_2O$ and/or sc-$H_2O$ reaction mixtures (Lu et al., 2013; Rangappa et al., 2008; Ziegler et al., 2001). Therefore, metal oxide particles with a stable surface modifier layer and narrow size distributions can be synthesized in an aqueous environment and subsequently transferred into nonpolar organic solvents where they often exhibit a very good dispersibility (Arita et al., 2010).

Both routes—the hydrothermal synthesis of functionalized and non-functionalized nanoparticle dispersions—were of interest for the work described in this thesis. A review of the pertinent literature shows that, despite the extensive amount of work already done on the continuous hydrothermal sysnthesis of metal oxide nanoparticles, the construction and operation of continuous pilot plants for this purpose still remains a considerable engineering challenge, a fact also reflected in the rather limited number of research groups working on this technology worldwide.

In order to effectively contribute to this field, a new pilot plant had to be designed and constructed that would be suitable for the synthesis of both non-functionalized and functionalized nanoparticles. Therefore, the pilot plant had to allow a great degree of flexibility in its configuration to accomodate different educt streams and in the geometry of both the reactor and necessary mixing unit(s). Furthermore, the pilot plant needed to accomodate a wide range of flow rates while at the same time ensuring good control of both temperature and pressure (very low pulsation).

To gauge the performance of the new pilot plant, a relatively simple and comparatively well known material system in continuous hydrothermal synthesis was selected: $CeO_2$ (Adschiri et al., 2000; Hakuta et al., 1998b; Zhang et al., 2007). $CeO_2$ nanoparticles have been studied extensively for applications in catalysis, sensors, electrochemistry, optics, biology and as an ultraviolet absorber and abrasive (Byrappa et al., 2008; Zhang et al., 2007). The structural, optical, electronic and catalytic properties of cerium oxide nanocrystals are strongly dependent on their geometrical parameters, especially size and shape. Therefore, the preparation of high quality ceria nanocrystals with desired morphology is of great fundamental and technological interest.

As a next step, the focus was on the continuous hydrothermal synthesis of iron oxide nanoparticles, a much more complex material system. We chose iron oxide as a model system for various reasons. First, iron oxide (nano)particles play an important role in natural processes and have manifold technical applications based on their special and diverse physical, chemical, and structural properties. Both structure and properties of iron oxide nanoparticles strongly depend on their particle size, their history of formation and the surrounding medium (Chernyshova et al., 2007). Second, the iron oxides/oxyhydroxides hematite, magnetite and ferrihydrite have been synthesized successfully in hot compressed water (Lam et al., 2008). Furthermore, recent studies for the evaluation of different mixers have used the synthesis of iron oxide nanoparticles from iron(III) nitrate as a model reaction, showing conversion rates higher than 96% for residence times as low a 100 ms and placing these results within the context of estimated $Fe_2O_3$ solubilities in near critical and supercritical water (Sue et al., 2011, 2010). Finally, it is possible to functionalize the surface of iron oxides in situ, opening the door to a wide variety of applications (e.g. Gupta and Gupta, 2005; Zhou et al.,

2011). For the reasons outlined above, the iron oxides are a material system with both a good basis of previous work to build upon and enough remaining challenges regarding their preparation, characterization and applications.

Unfortunately, the characterization of iron oxides is complicated by the existence of (at least) 16 different types, with transitions possible between some of them depending on the processing steps before or even during characterization. In general, of the many published reports on the hydrothermal synthesis of metal oxide nanoparticles, most have focused on characterizing the particles obtained after drying (which can change the structure and properties of the metal oxide nanoparticles) and only in a few reports the properties and particle size distributions of the particles were examined in the dispersed form in which they are obtained through hydrothermal synthesis (e.g. Hayashi et al., 2010a,b, 2009; Noguchi et al., 2008; Tyrsted et al., 2010). Insofar as the iron oxide nanoparticles are destined to be used in dispersed form, the characterization of the as-obtained dispersions is important and was undertaken in the present work. In addition, investigations of the as-obtained dispersions can be helpful in drawing conclusions about the mixing step—most often a key factor both in reproducibility and in obtaining narrow size distributions.

For the characterization of the size distribution of the dispersed particles with high resolution and reproducibility, we turned to differential centrifugal sedimentation analysis (see e.g. Bell et al., 2012, for a comparison with alternative techniques). This technique has proven to be a powerful and convenient tool e.g. for examining nanocomposite dispersions, for abrasive particles in CMP (chemical-mechanical planarization slurries) and colloidal carbon nanotube dispersions (Balmer et al., 2010; Kamiti et al., 2007; Nadler et al., 2008). Its strength is in the resolution of closely spaced subpopulations (Bell et al., 2012), an issue of great importance when examining the effect of process—and especially mixing—conditions on the resulting particles.

To characterize the structure of the iron oxide nanoparticles, we turned to Raman spectroscopy using a newly developed measurement setup for performing Raman spectroscopic investigations directly in the aqueous phase. Though the Raman technique has been applied to the study of iron oxides since the 1970s (Beattie and Gilson, 1970; Reid et al., 1977; Thibeau et al., 1978), the co-existence of different ferric oxide and oxyhydroxide phases, the sometimes non-stoichiometric composition of the iron oxides and oxyhydroxides as well as laser-induced thermal effects and phase changes remain challenges in interpreting recorded Raman spectra correctly (Bersani et al., 1999). More recently, a few studies have reported on detailed investigations with synthetic and natural iron oxide and oxyhydoxides revealing that Raman spectroscopy has the potential to rapidly dinstinguish between the numerous iron oxides and oxyhydroxides when sufficiently low laser powers are applied to avoid phase transformation upon laser irradiation (Chernyshova et al., 2007; Hanesch, 2009; Jubb and Allen, 2010). These recent reports were the starting point for the investigations by means of Raman spectroscopy in the context of this thesis.

Therefore, by combining the use of a highly flexible pilot plant for the hydrothermal nanoparticle synthesis with suitable ex situ methods for characterizing the as-obtained nanoparticles in dispersion, it is possible to make a contribution to the screening for favorable process conditions to obtain stable iron oxide nanoparticle dispersions beyond what is already known from previous studies (e.g. Aizawa et al., 2007; Blood et al., 2004; Bremholm et al., 2009a; Kawasaki et al., 2010; Lester et al., 2006; Mae et al., 2007; Middelkoop et al., 2009; Sue et al., 2010; Tyrsted et al., 2010). In particular, we investigated the effect that changing the ratio of the cold and hot streams, the fluid velocity and the identity of the educt (see Figure 1.2) has on the size distribution of both primary particles and particle agglomerates, on the colloidal stability of the dispersions and on the structure of the particles measured directly in the as-produced aqueous dispersions.

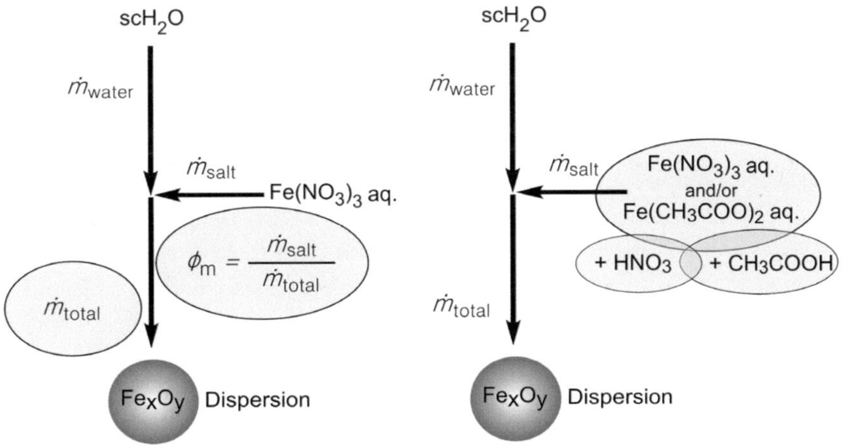


**Figure 1.2:** Schemes of the flow configuration in the core part of the pilot plant; (left) for the study on varying the flow conditions; (right) for the study on varying the chemical compositions of the metal salt stream.

The work outlined above was used as the basis for establishing a method to produce functionalized iron oxide nanoparticles. These are an example of core-shell nanoparticles consisting of an inorganic core and an organic shell of a diverse range of molecules which represent an important class of organic-inorganic hybrid materials. Hybrid organic-inorganic materials span a very wide range of known, envisioned and yet-to-be discovered properties with potential application e.g. in various technologies related to water and energy supply and efficiency, environmental protection, health, microelectronics, optics and transportation. In the particular case of functionalized metal oxide nanoparticles, the metal oxide nanocrystal cores are useful because the properties can easily be tuned by adjusting the crystal composition, size and shape. To benefit from the complementarities and synergies between such materials and the specific properties and functions of the organic shell, it is necessary to develop a convenient process for the synthesis of metal oxide nanoparticles with universal and accessible anchors on their surface (Beija et al., 2011; Sanchez et al., 2011; Yin and Alivisatos, 2005).

The continuous hydothermal synthesis of metal oxide nanoparticles can be modified to produce organic-inorganic hybrid particles by in situ functionalization with organic molecules such as thiols, alcohols, aldehydes, carboxylic acids or amines (Mousavand et al., 2007; Ziegler et al., 2001) as organic modifiers in homogeneous nc-$H_2O$ and/or sc-$H_2O$ reaction mixtures (Rangappa et al., 2008; Ziegler et al., 2001). It has been shown that metal oxide particles with a stable surface modifier layer and narrow size distributions can be synthesized in an aqueous environment which can be transferred into nonpolar organic solvents where they often exhibit a very good dispersibility (Arita et al., 2010). We sought to extended this concept to produce functionalized metal oxide nanoparticles with terminal groups suitable for further modification by *click* chemistry. By definition, *click* reactions are simple, efficient and applicable in diverse modular approaches without the formation of significant side products (Kolb et al., 2001), making *click* chemistry a promising tool in colloidal science (Lutz and Zarafshani, 2008) and, increasingly, in chemical and biological engineering (Adzima and Bowman, 2012). The design of bifunctional molecules that combine a robust anchor that can bind to a variety of metal oxide surfaces and a second functional group that leads to covalent attachment of molecules via *click* chemistry has been proposed as a strategy to introduce a broad range of functionalities onto the surfaces of metal oxide nanoparticles (White et al., 2007). The strategy is applicable to a tremendous diversity of molecules including functional polymers, biomolecules, fluorescent dyes and molecular anchors. However, to the best of our knowledge, all *clickable* metal oxide nanoparticles have been synthesized thus far using methods that require multiple steps, long reaction times and/or organic solvents (e.g. Goldmann et al., 2010; Li and Binder, 2011; White et al., 2007). Hence, there is a need for straightforward, single-step procedures for the versatile synthesis and functionalization of nanoparticles, preferably in an aqueous environment.

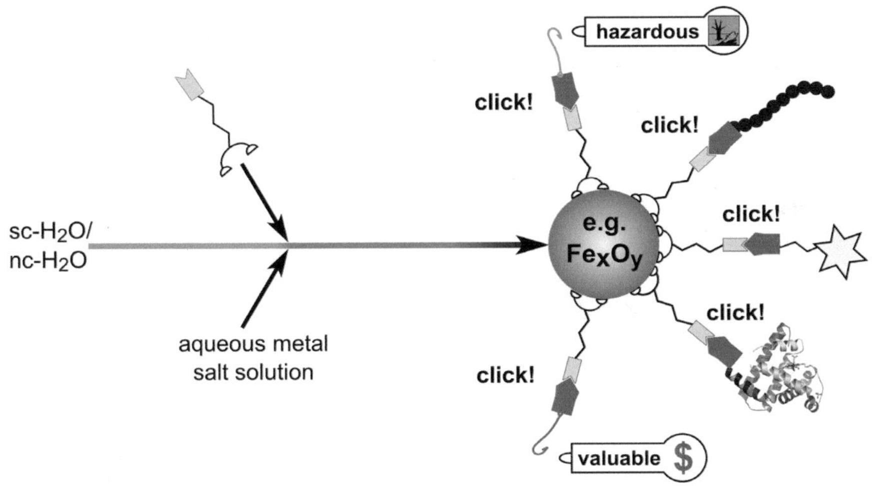


**Figure 1.3:** Schematic representation of the process for the synthesis of metal oxide nanoparticles presenting *clickable* anchors, and selected potential applications.

An attempt was therefore made to introduce organic molecules onto the surface of the nanoparticles comprising two different functional groups, one forming a stable bond with the metal oxide surface during the hydrothermal process and the other presented as an accessible reactive group for secondary functionalization, as shown conceptually in Figure 1.3. For this purpose, molecules were chosen that contain a terminal alkyne group allowing straightforward coupling with azide-terminated molecules via the most used and well-known *click* reaction, the copper(I)-catalyzed azide-alkyne cycloaddition (CuAAC) (Kolb et al., 2001). Versatile and facile methods for the preparation of azide-terminated polymers and other molecules have been described (Lu et al., 2009). The use of iron oxide nanoparticles as inorganic cores offers the advantage of biocompatibility and low toxicity (Ciampi et al., 2011; Gupta and Gupta, 2005), which has led to significant research into their use (as nanoparticles) in diagnostics and medical technology (Teja and Koh, 2009). Example applications are their use as contrast agents for in vivo magnetic resonance imaging (MRI) in molecular and cell biolgy, hyperthermia inducing agents in cancer therapy (Hayashi et al., 2010a) as well as in cell separation, organic and biochemical syntheses, protein purification, for capturing toxic metals from liquid matrices for industrial applications as well as environmental remediation (Rutledge et al., 2010; Teja and Koh, 2009).

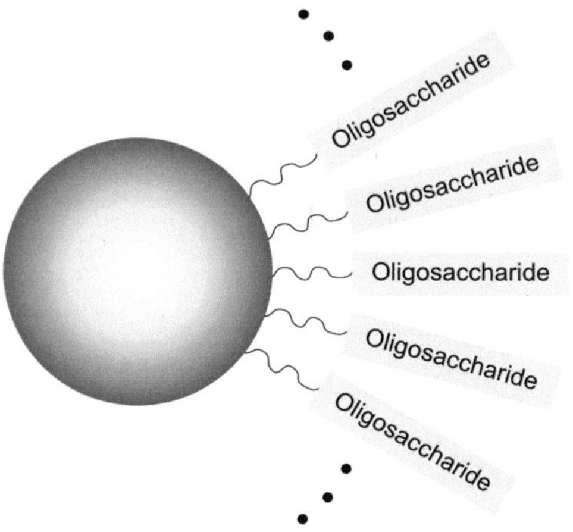


**Figure 1.4:** Schematic representation of a nanoparticle presenting clustered oligosaccharides for exploiting the "cluster glycoside effect".

A very attractive and increasingly important kind of functionalization of nanoparticles are complex carbohydrate structures, the so-called oligosaccharides, which when for example bound to nanoparticles can mimick the "glycocalyx" covering the surface of mammalian cells. On the cell surface, the oligosaccharides are covalently bound to proteins and lipids (together called glycans) and perform their various biological functions (including cell communication and trafficking, immune responses and tumor genesis and progression) through the adoption of different three-dimensional structures (de la Fuente and Penades, 2006; Sihelníková and Tvaroška, 2007; Taylor and Drickamer, 2003; Wang et al., 2010). In performing these functions, extremely weak carbohydrate–carbohydrate and carbohydrate–protein interactions play a central role (Sihelníková and Tvaroška, 2007). These interactions are amplified by the clustering of ligand and receptors to form microdomains leading to the "cluster glycoside effect", thus enabling their biological significance. Both the weakness of the interactions underlying their biological function as well as the synthetic difficulty in preparing glycans are responsible for their underestimation in the past as compared to peptides and nucleotides. However, recent advances in carbohydrate synthesis and glycan analysis have led to a significant increase in research efforts and publications in this field (Lepenies et al., 2010; Wang et al., 2010). One possible way of exploiting the "cluster glycoside effect" for different applications is the preparation of nanoparticles covered with a dense coat of oligosaccharides (see Figure 1.4). Potential applications of glyconanoparticles include the use as anti-cancer agents (e.g. Cipolla et al., 2008), anti-viral agents (e.g. against HIV: Martínez-Ávila et al., 2009), vaccines (e.g. anticancer: Ojeda et al., 2007), diagnosis (e.g. pre-symptomatic in vivo imaging of brain diseases such as multiple sclerosis (MS), ischemic stroke etc.: Serres et al., 2009; van Kasteren et al., 2009), and bacterium detection and decontamination (El-Boubbou et al., 2007). With this enormous potential in mind, we aimed to synthesize model glyconanoparticles on the basis of the in situ functionalized (*clickable*) iron oxide nanoparticles using a simple saccharide (galactose) with an azide linker.

Altogether, this thesis:

- Describes the design, construction and testing of a flexible pilot plant for the continuous hydrothermal synthesis of non-functionalized and functionalized metal oxide nanoparticles under near-critical and supercritical conditions
- Provides insights into the effect of parameter changes on the product properties, including detailed analysis of the as-obtained dispersions
- Expands the spectrum of possibilities for synthesizing in situ functionalized metal oxide nanoparticles via a continuous hydrothermal process
- Demonstrates a general strategy for the preparation of functionalized metal oxide nanoparticles on the basis of *click* chemistry
- Shows a simple route for the preparation of glyconanoparticles.

# Chapter 2

# Background

## 2.1 Hydrothermal synthesis under near-critical and supercritical conditions

Interest in the properties of water near and above the critical point ($T_c$ = 647.1 K, $p_c$ = 22.1 MPa, $\rho_c$ = 320 kg m$^{-3}$)* was for a long time mostly related to its use in power plants (high-pressure steam) and naturally-occurring hydrothermal reactions. Hot compressed aqueous solutions have played a crucial role in the development of the earth e.g. by creating mineral deposits and soils, and are involved in the formation of natural gas (Bondarenko et al., 2006; Weingärtner and Franck, 2005). Starting in the late 1970s, interest in the commercial use of supercritical water technology shifted towards alternative fuels, coal and biomass conversion, and waste treatment as a result of the first oil crisis and environmental concerns (Brunner, 2009; Eckert et al., 1996).

The main focus of investigation has been the use of hot compressed water as a "green" solvent: It is cheap, non-toxic, non-flammable, and non-explosive, and the lack of interfacial mass transfer resistance combined with high reaction temperatures leads to short residence times and high space yields (Prikopsky et al., 2007; Rogalinski et al., 2008). Moreover, the properties of hot compressed water are tunable by adjusting temperature and pressure. The push for industrial applications of near-critical and supercritical water was first especially related to its use for hazardous waste destruction. The supercritical water oxidation (SCWO) allows the complete oxidation of hazardous organic wastes without the need for post-treatment (Brunner, 2009; Marrone

*Hot compressed water, hydorthermal and water under near-critical and supercritical conditions are used largely interchangeably in this work.

et al., 2004; Portella et al., 2007). Further applications are supercritical water gasification (SCWG) where e.g. wood-based biomass is converted to $CH_4$, $H_2$ and $CO_2$ (Peterson et al., 2008), and hydrothermal processing of biomass-derived chemicals such as sugars, starch and cellulose which can be converted to platform chemicals such as polyalcohols and hydroxymethylfurfural and ultimately to biobased materials (e.g. polymers; Hirth et al., 2006). Finally, the use of near-critical and supercritical water in material synthesis (such as nanoparticle synthesis, discussed below) has gained significant momentum (Aymonier et al., 2006). Nevertheless, only a limited number of practical applications of supercritical water processes have so far been realized on an industrial scale due to difficulties caused by the demanding operating conditions—implying high equipment costs—combined with the properties of water and the species involved, particularly precipitating salts (Bicker et al., 2005; Brunner, 2009; Butenhoff et al., 1996).

The following sections summarize important findings on the properties of supercritical water from both a macroscopic and microscopic perspective, and provide a brief summary of the use of suprecritical water as a medium for the preparation of metal oxide nanoparticles and their functionalization.

### 2.1.1 The properties of water

The properties of water drastically change around the critical point where the properties of the gaseous and liquid phase become identical (see Figure 2.1, left). The well known unusal properties of water under ambient conditions are mostly related to the hydrogen bond and the tetrahedral coordination that the water molecules adopt (see Figure 2.1, right). The tetrahedral coordination of water and the hydrogen bonds remain under near-critical and supercritical conditions, although the resulting networks are confined to a shorter range (Weingärtner and Franck, 2005), further discussion see below on the section on the molecular perspective. In addition, the density of water changes strongly with changes in both temperature and pressure in the vicinity of the critial point (Jähnke, 2000), as illustrated in Figure 2.2 (top left). Together with the density, the dielectric constant of water also experiences drastic changes around the critical point (Figure 2.2, top right). The reduction in dielectric constant seen between room temperature and pressure ($\epsilon \approx 80$) and the critical point ($\epsilon \approx 6$) has important consequences for the water solubility of compounds (Deul and Franck, 1991), e.g.:

- The solubility of inorganic salts is strongly reduced around the critical point compared to room temperature and pressure (Thomason and Modell, 1984)
- The chemical equilibrium of solvated metal ions shifts toward the corresponding metal oxide or hydroxide around the critical point (Adschiri et al., 1992)
- The solubility of nonpolar compounds is high under near and supercritical conditions compared to room temperature and pressure (Kruse and Dinjus, 2007)
- The solubility of gases such as oxygen, nitrogen, hydrogen or carbon dioxide is essentially unlimited (miscible) (Adschiri et al., 1992).

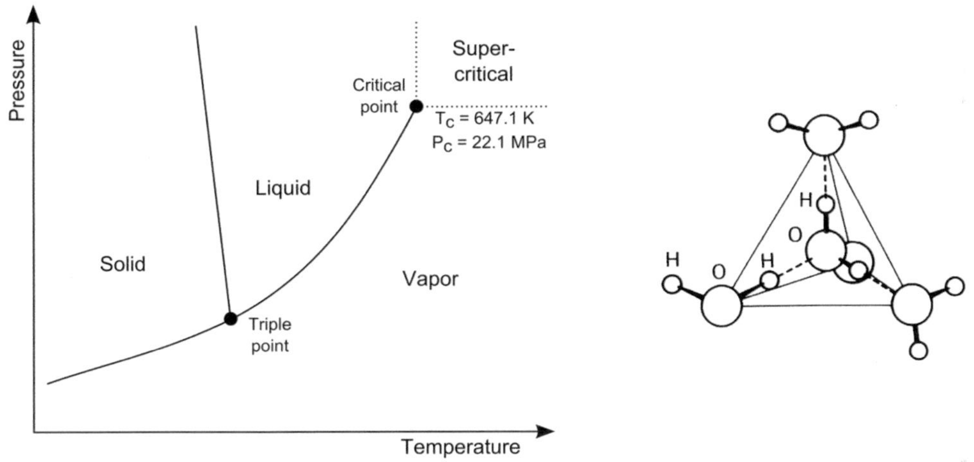


**Figure 2.1:** (left) Vapor pressure curve of water (after Van Wylen et al., 1994); (right) scheme of water molecules adopting a tetrahedral coordination by hydrogen bonds taken from Israelachvili (1992).

A further property that changes strongly in the vicinity of the critical point is the self-dissociation of water: when approaching the critical point, the self-dissociation of water increases one to three orders of magnitude compared to ambient conditions but decreases by several orders of magnitude above the critical point, as shown in Figure 2.2 (bottom left; Marshall and Franck, 1981). These changes have an enormous influence on both hydrolysis and acid-base equilibria (Weingärtner and Franck, 2005) and therefore on the solubility and chemical equilibrium of compounds formed by metal ions in water (Adschiri et al., 2001).

In addition, the key transport property—viscosity—experiences a strong decrease with temperature and this decrease is particularly steep around the critical point (see

Figure 2.2, bottom right). For the supercritical conditions relevant to this work, the viscosity is approximately two orders of magnitude lower than that of water under ambient conditions and therefore typically less than one order of magnitude higher than that of gases. At the same time, the density of supercritical water—1/2 to 1/10th that of water at room temperature and pressure—is approximately two orders of magnitude higher than for typical gases. Consequently, mass transport processes are accelerated under supercritical conditions, favoring mass transport limited reactions—a property that makes supercritical water a particularly attractive medium for chemical processes (Eckert et al., 1996; Kruse and Dinjus, 2007; Weingärtner and Franck, 2005).

Besides the knowledge of its bulk properties, the molecular perspective is of great importance for the understanding of the hot compressed water environment due to the microheterogenous distribution of water under near- and supercritical conditions (Balbuena et al., 1994; Kruse and Dinjus, 2007). This microheterogeneity arises because, on the average, water molecules are bonded to each other almost half as strong under supercritical conditions as compared to ambient water due to the decreased density and loss of tetrahedral coordination. However, both measurements and simulations have shown that water molecules connect through hydrogen bonds under supercritical conditions, e.g. forming clusters of mostly five or fewer molecules at 773 K (but clusters can contain up to 20 or more molecules because the distribution is very broad). The formation of these clusters leads to local differences in density of hot compressed water, which are rather extreme in water near the critical point. The breakdown of the long range networks of water molecules connected by hydrogen bonds, combined with a shorter lifetime of the individual hydrogen bonds in hot compressed water, leads to decreased energetic barriers for rotation and translation of the individual water molecules, which in turn results in increased self-diffusivity and enhanced heat transport (Kruse and Dinjus, 2007).

The formation of large clusters near the critical point leads to locally increased concentrations of $H^+$ and $OH^-$ ions and has been reported to accelerate certain reactions. In addition, both molecular simulations and measurements have shown that the density of hot compressed water is higher in the vicinity of ions than in the bulk fluid as a result of a largely intact primary solvation shell for both monovalent and divalent cations and a partly intact second solvation shell for divalent cations. For anions, the effect is of a similar nature but of less magnitude because the first solvation shell is only weakly preserved. This increased solvent density around ions affects both their

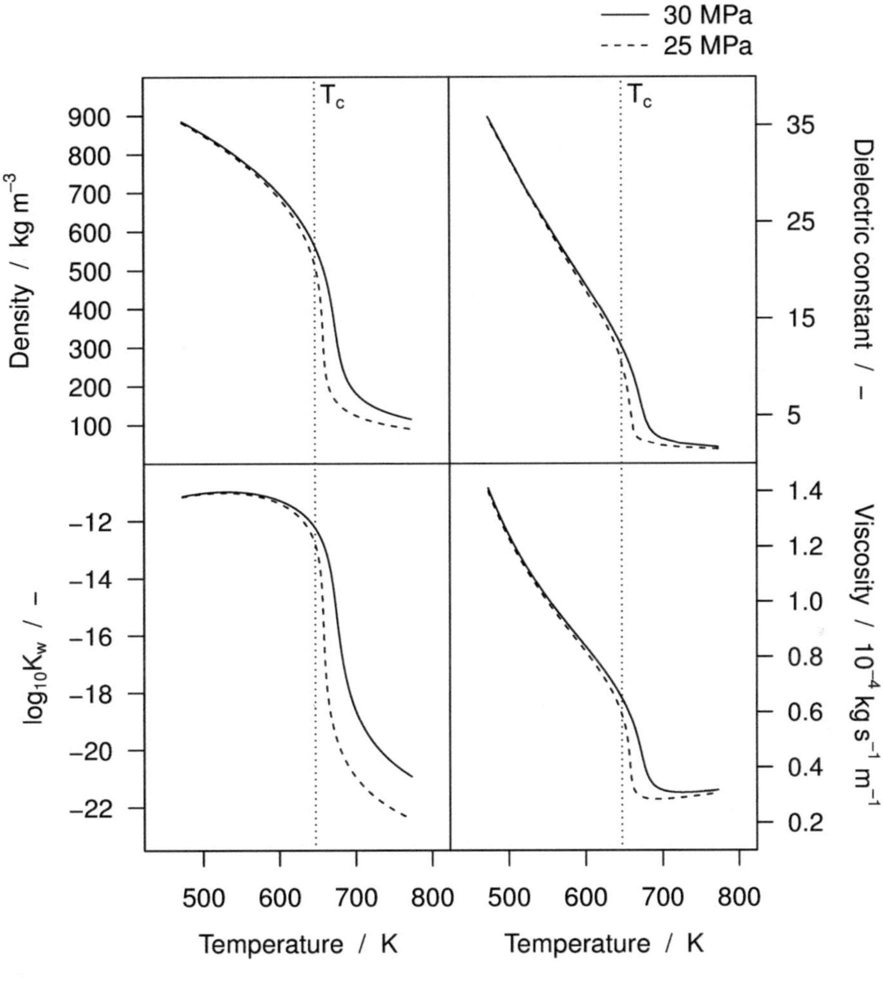


**Figure 2.2:** Important properties of water in the relevant parameter range used within this work. (top left) Density (from the formulation of IAPWS 2001, Wagner and Pruss, 2002); (top right) dielectric constant (data from Heger et al., 1980; Johnson and Norton, 1991); (bottom left) ion product (data from Marshall and Franck, 1981); (bottom right) viscosity (data from IAPWS, 1997; Kestin et al., 1984).

chemical potential and solvation energy, and is usually taken into account in calculations by applying a "local relative dielectric constant" higher than the bulk. Unlike in water at ambient conditions, salts like NaCl, LiCl, ZnBr are dissolved as ion pairs and even higher clusters in hot compressed water, with the degree of dissociation increasing with decreasing concentration of salts and increasing density of the hot compressed water. This combination of a polar molecule creating a polar microenvironment but behaving macroscopically like a non-polar solvent opens up possibilies for unconventional reactions (Kruse and Dinjus, 2007). This as well as the exploitation of the drastic changes in the properties from ambient to near-critical/supercritical conditions form the basis of the material synthesis described below.

### 2.1.2 The continuous hydrothermal synthesis of metal oxide nanoparticles

The synthesis of inorganic crystals under hydrothermal conditions was carried out as early as the 1840s, but with the objective of preparing comparatively large particles of several millimiters in size. However, the resulting particle dispersions were often discarded as failures because the particles were too small and could not be analyzed with the methods available at that time. With the development of particle analysis techniques based on X-rays and the ability to collect diffraction patterns from fine powders, hydrothermally synthesized metal oxide nanoparticles started to be characterized in the 1920s-1950s, and even in more detail starting in the 1980s with the advent of high-resolution scanning electron microscopy (SEM). Starting in the 1990s and continuing until today, work has turned towards producing particles with a certain degree of control over their size and morphology in order to control their properties (Byrappa and Adschiri, 2007).

The physico-chemical basis for the hydrothermal synthesis of metal oxide nanoparticles is the change in the properties of near-critical and supercritical water compared to water under ambient conditions. Of particular importance is the decrease in the dielectric constant (discussed above and shown in Figure 2.2) which affects the solubility of metal ions (Hakuta et al., 2001). Simplified, the synthesis of the metal oxides occurs in two consecutive (equilibrium) reaction steps (Adschiri and Arai, 2002):

1. Hydrolysis of the metal ions and formation of metal hydroxides.
2. Dehydration of the metal hydroxides to metal oxides and water.

With increasing temperature, the equilibria of compressed aqueous metal salt solutions shift in the direction of the metal (hydr)oxides (Adschiri et al., 2001, 1992). Moreover, because the solubility of metal ions decreases by several orders of magnitude around the critical point, a switch from ambient to near-critical or supercritical conditions leads to high supersaturations resulting in high nucleation rates and the formation of very fine particles, as shown in Figure 2.3 (Hakuta et al., 2001). Based upon this principle, both batch and continuous processess have been developed for the hydrothermal synthesis of metal oxide nanoparticles.

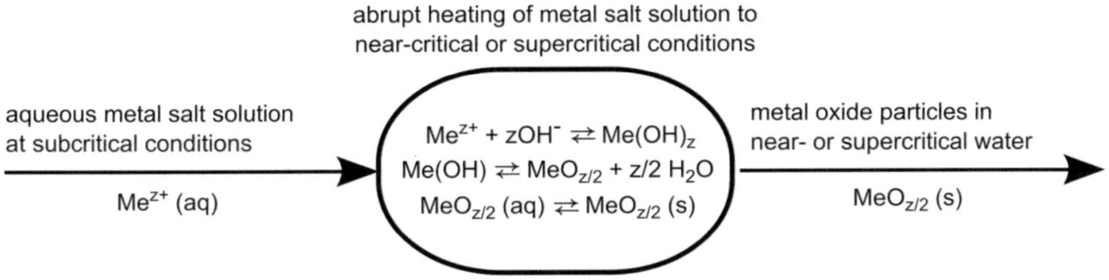


**Figure 2.3:** Schematic of an ideal hydrothermal process. The reactions are simplified for clarity.

The hydrothermal synthesis of metal oxide nanoparticles has been typically carried out as a batch process using autoclaves: The aqueous solutions are heated up to 373–573 K and are then aged for several hours or days (Adschiri et al., 2001). This process has the advantage of modest equipment requirements, but because the heating of the aqueous metal solutions in autoclaves occurs slowly, the result are continuously changing equilibrium conditions and very long reaction times (Adschiri and Arai, 2002).

In the continuous hydrothermal process, the aim is to effect the required change in temperature (to near-critical or supercritical conditions) very rapidly through the mixing of the compressed cold metal salt stream with a sufficiently hot compressed water stream. A feature of this approach is that, as a result of the faster and more homogenous temperature change, particle formation is faster and generally leads to more uniform particle size distributions and smaller particles (Adschiri et al., 2001). Because of the strong changes in properties of water in the vicinity of the critical point (see Figure 2.2), it is possible to control the reaction conditions through variations in pressure and temperature, which in turn allows the tuning of the resulting particle sizes and their distribution, as well as morphology and structure of the particles (Adschiri et al., 2001; Lam et al., 2008). In turn, the sensitivity of the process to changes in temperature and pressure implies high requirements for the measurement and control of these two parameters. A scheme of a typcial hydothermal flow type reactor is shown in Figure 2.4. The continuous process has already been successfully implemented to produce a wide variety of metal and mixed-metal oxides including $CeO_2$, $ZnO_2$, $\alpha$-$Fe_2O_3$, $Fe_3O_4$, $Co_3O_4$, $CoFe_2O_4$, AlOOH, $BaTiO_3$, $ZrO_2$, $TiO_2$, $LiCoO_2$, $CuMn_2O_4$, $ZnGa_2O_4$:$Mn^{2+}$, and $Ca_{0.8}Sr_{0.2}Ti_{1-x}Fe_xO_{3-\delta}$ ($x = 0.1$–$0.3$) (e.g Adschiri et al., 1992; Cabanas and Poliakoff, 2001; Cote et al., 2003; Hakuta et al., 1998a, 2004, 1998b, 2005; Hao and Teja, 2003; Hayashi et al., 2013; Lu et al., 2008; Sue et al., 2003, 2006). Furthermore, due to the advancements in automization, process control, and data processing, it has become possible to systematically explore whole phase diagrams of mixed metal oxides in the search for specific properties (high-throughput screening, e.g. Kellici et al., 2010; Lin et al., 2010; Weng et al., 2009).

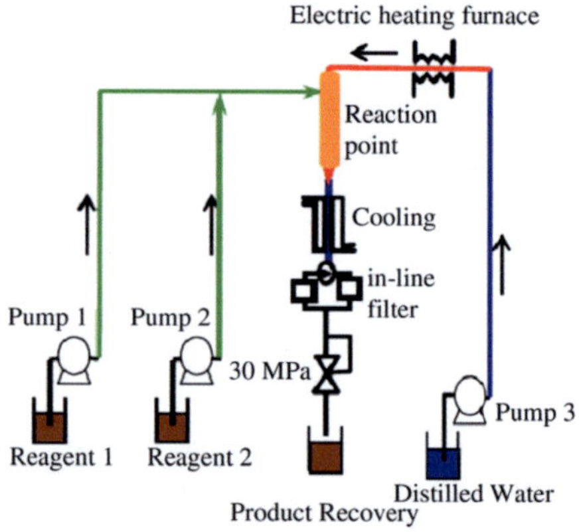


**Figure 2.4:** Schematic of a typical hydrothermal flow type apparatus (from Byrappa et al., 2008) and used with permission from Elsevier.

In addition to demonstrating the synthesis of diverse metal and mixed-metal oxides, considerable effort has been invested in elucidating the influence of the process conditions and the starting materials on the product properties such as particle size distribution and morphology. The synthesized particles have been extensively characterized as dry powders. However, only few reports exist were the properties of the particles were examined in dispersed form (e.g. Hayashi et al., 2010a,b, 2009; Noguchi et al., 2008; Tyrsted et al., 2010).

The observation that different material systems react differently to parameter changes in the hydrothermal synthesis of inorganic particles has triggered interest in an improved theoretical understanding of this process. Theoretical studies are important for guiding experimental investigations into the tuning of the properties of the resulting particle system and provide insights not or at least not readily obtainable from experiment.

The knowledge of the chemical equilbria of the reactions involved forms the basis for the understanding and prediction of the formation of a certain or desired product. Based on the revised Helgeson-Kirkham-Flowers model (HKF, Shock et al., 1992), which has been used to calculate equilibrium constants and solubilities for hundreds of inorganic species in the temperature range $\approx$ 300–1250 K and pressure range $\approx$ 0.1–500 MPa, simplified equations have been proposed eliminating the need for the partial molar volume, which is often not available under these conditions in the literature (Adschiri et al., 2001). It was shown that even the simple equation shown below could often estimate the equilibrium constant appropriately:

$$\ln K(T,\rho) = \ln K(T_0,\rho_0) + \frac{\Delta H}{R}\left(\frac{1}{T} - \frac{1}{T_0}\right) \underbrace{- \frac{\omega_j}{RT}\underbrace{\left(\frac{1}{\epsilon(T,\rho)} - \frac{1}{\epsilon(T_0,\rho_0)}\right)}_{\text{dielectric factor (see Figure 2.5)}}}_{\text{Born term}} \tag{2.1}$$

where $K$ is the equilibrium constant, $T$ is the absolute temperature, $T_0$ is the room temperature, $R$ is the gas constant, $\epsilon$ is the dielectric constant, $\omega_j$ is a parameter determined by the reaction system, $\rho$ is the density, and $\rho_0$ is the density at ambient conditions (Adschiri and Arai, 2002). The so called Born term expresses the change in the electrostatic interaction of an ion with its surrounding medium when moving between media of different dielectric properties. Thus, the magnitude of this term is dictated by the difference of the inverse of the dielectric constant between the two media as captured in the "dielectric factor". Because the dielectric constant of water is very different under near-critical and supercritical conditions compared to ambient conditions, the dielectric factor increases strongly around the critical point (see Figure 2.5). However, the change in the dielectric factor is more abrupt than that for the dielectric constant around the critical point (cf. Figure 2.2). It is important to note that this macroscopic approach (using the bulk dielectric constant of water) does not reflect the microheterogeneity of near-critical and supercritial water (see Section 2.1.1).

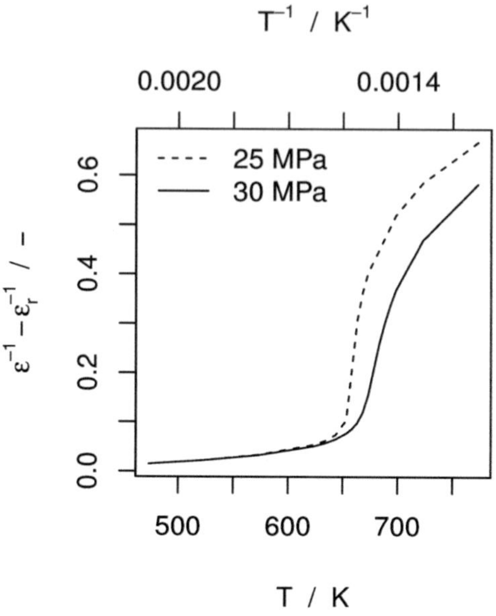


**Figure 2.5:** Dielectric factor around the critical point (data from Heger et al., 1980; Johnson and Norton, 1991). Reference point: $\epsilon_r = 78$.

Using this approach, the concentration of dissolved species can be estimated by solving the system of nonlinear equations of the relevant equilibrium constants, mass balance relations, the charge balance relation, and formulation for the ion activity coefficients. Recalling that the metal oxide solubility is defined as the total concentration of the metal ion aqueous complex species, it is possible to estimate the supersaturation—the driving force for precipitation—as:

$$\sigma = \frac{S_0 - S}{S} \tag{2.2}$$

where $S$ denotes the estimated metal oxide solubility under near-critical or supercritical conditions and $S_0$ denotes the metal concentration of the starting solutions (Sue et al., 2006). In a study by Sue et al. (2006), different metal oxide nanoparticles were synthesized under the similar process conditions (T = 673 K, p = 25–37.5 MPa, $\tau$ = 1 s) and the resulting particle sizes (from TEM and PXRD analysis) were placed in the context of the estimated supersaturation (see Figure 2.6). As expected from nucleation theory, the average particle size of the synthesized metal oxides tends to decrease with increasing supersaturation (Sue et al., 2006). However, the results also show that for the precise control and tuning of the particle size, other underlying factors have to be considered.

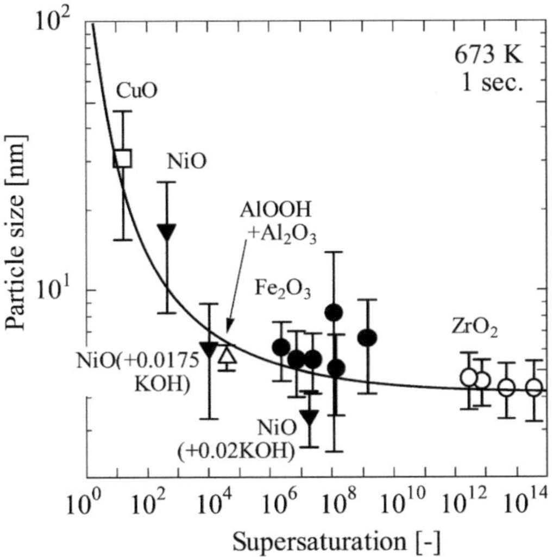


**Figure 2.6:** Average particle size of metal oxide nanoparticles synthesized hydrothermally in a flow through apparatus at pressures between 25 MPa and 38 MPa (figure taken from Sue et al., 2006). Note that the curve is only a guide to the eye. Reproduced by permission of The Royal Society of Chemistry.

This "simple" approach for the estimation of the supersaturation assumes infinitely fast reaction kinetics and mixing. Whereas considerable effort has been undertaken to understand, investigate and control the mixing between the cold metal salt stream and hot compressed water stream, as far as we know, fairly little has been investigated and is known on the kinetics of particle formation, especially in the early stages. It is out of question that the reaction rates also change greatly in the region around the critical point. Simple equations have been proposed including a Born type term (Adschiri et al., 2001):

$$\ln k = \ln a_r + \frac{E_\mathrm{r}}{RT} + \frac{\Psi}{RT}\left(\frac{1}{\epsilon} - \frac{1}{\epsilon_\mathrm{r}}\right) \tag{2.3}$$

where $a_\mathrm{r}$ and $E_\mathrm{r}$ are a frequency factor and activation energy, respectively, of the reaction at dielectric constant $\epsilon_\mathrm{r}$, and $\Psi$ is a constant determined by the reaction system.

However, here the microheterogeinity of the water under near-critical and supercritical conditions is again not taken into account. For a few material systems, the overall reaction rates have been investigated by determining the conversion ($X$) from the concentration of the respective metal ion in the effluent after different residence times for different reaction temperatures. For the starting material $Al(NO_3)_3$, the mathematically transformed conversion as a function of the residence time—assuming a first order reaction type—is shown in Figure 2.7 (left) for different reaction temperatures. First order reaction rate constants can be obtained from this type of plot. A selection of Arrhenius plots for the first order reaction rate contstants is depicted for different material systems in Figure 2.7 (right).

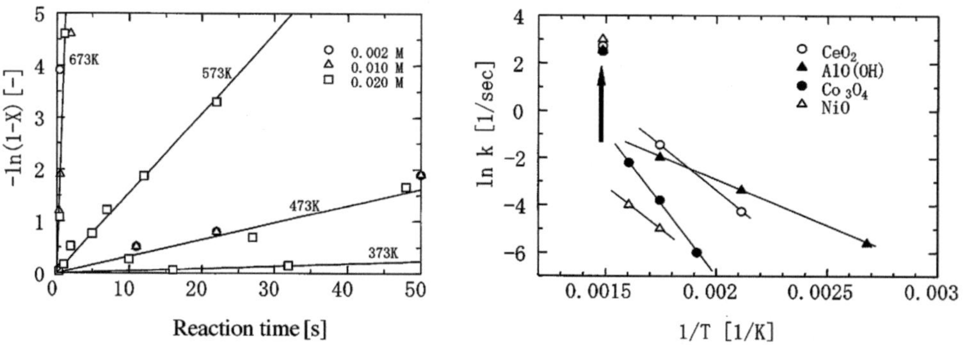


**Figure 2.7:** (left) Conversion (transformed) as a function of residence time for different reaction temperatures of a continuous hydrothermal process using $Al(NO_3)_3$ as starting material. (right) Arrhenius plot of first order rate constants estimated from the slopes of the plots of $-\ln(1-X)$ vs. residence time for different material systems (both figures taken from Adschiri et al., 2001, and used with permission from Springer).

The kinetics of particle formation in hydrothermal processes has also been studied by means of molecular simulations. In particular, using the (non-reacting) NaCl system, it has been reported that particle formation occurs within a few hundred picoseconds after the jump from ambient to supercritical conditions (Luemmen and Kvamme, 2007; Nahtigal et al., 2008). However, these investigations are based on an instantaneous change of conditions. In reality, the change from ambient to near-critical or supercritical conditions as well as the formation of the metal (hydro)oxides require a finite amount of time.

A different approach for a more in depth understanding of the process is in situ investigations of particle nucleation and growth by means of X-ray techniques using custom-made equipment and high-intensity synchrotrons (e.g. Bremholm et al., 2008). Using a combination of SAXS and WAXS (performed simultaneously) with $Ce_xZr_{1-x}O_2$ ($x$ = 0.0, 0.2, 0.5, 0.8, 1.0) as a material system, it was shown that particle growth can be very different at the mechanistic level (diffusion vs. reaction limited) for different materials (Ce-rich vs. Zr-rich) and that the growth mechanism can change from diffusion-limited to reaction-limited or vice versa depending on the size of the particles for a given composition (Tyrsted et al., 2010). However, in situ studies have been generally limited to either batch processes (e.g. Tyrsted et al., 2010) or, in the case of continuous processes (e.g. Bremholm et al., 2008), to sections of the reactor a certain distance away from the point where the cold metal salt stream is mixed with the hot (supercritical) water stream (cf. Figure 2.4). In the latter case, this means that in situ investigations have not provided insights into the crucial initial moments of particle nucleation and growth in continuous hydrothermal synthesis, where heating and mixing play a central role.

In a real continuous flow setup, the rapid change from ambient to near-critical or supercritical conditions is achieved through the mixing of a compressed aqueous stream containing the metal salts and a sufficiently hot water stream. To accomplish the mixing, high pressure mixing units of different geometries have been used. Already early studies implicitly or explicitly referred to problems with plugging in and around the mixing unit and very broad particle size distributions, indicating that achieving sufficiently fast and homogenous mixing is essential to the success of hydrothermal flow synthesis if it was to become a reliable process for the production of metal oxide nanoparticles (Blood et al., 2004; Bremholm et al., 2008; Lester et al., 2010, and citations therein). In this context, "fast" is ideally much faster than the chemical reactions, nucleation and formation of particles, a process thought to require 10–100 ms. In addition, homogeneous mixing is required because stagnant zones in the mixing unit will lead to an accumulation of particles, in turn leading to non uniform mixing, a poorly controlled product and eventually to blocking of the apparatus (Lester et al., 2012). This prompted several studies on the mixing step including CFD simulations (e.g. Kawasaki et al., 2010; Sierra-Pallares et al., 2011; Sue et al., 2010), experimental studies with different mixing units in continuous hydrothermal synthesis apparatuses (e.g. Hong et al., 2013; Kawasaki et al., 2009; Sue et al., 2010; Wakashima et al., 2007), and in situ measurements using a dye in model systems under ambient conditions or supercritical conditions (e.g. Aizawa et al., 2007; Blood et al., 2004; Lester et al., 2012).

All studies into the mixing step in continuous hydrothermal synthesis face the same set of difficulties: on the one hand, the mixing behavior between the metal salt stream and the hot compressed water stream is unusually complex due to the large differences in temperature, density and viscosity between the two fluids, such that buoyancy can be an important driver for mixing in hydrothermal mixers/reactors (Lester et al., 2012). On the other hand, making direct measurements in the mixing unit is extremely difficult so that most studies approach the problem by means of indirect observations. For example, Kawasaki et al. (2010) showed that the diameter of NiO nanoparticles synthesized continuously under hydrothermal conditions (673 K, 30 MPa) was reduced upon decreasing the inner diameter of the mixing unit (T-geometry) or increasing the flow rate to increase turbulence within the mixing unit. They combined their experimental results with CFD simulations and concluded that the observed effects correlate well with the heating rate of the metal salt solutions, variable through variations in the mixing rate (see Figure 2.8). Other studies, such as that of Lester et al. (2012), have reported results based on model mixing systems exhibiting analogous properties under ambient conditions (similar ratio of inertial to viscous forces).

Despite the difficulties in making direct measurements of a mixing unit, Aizawa et al. (2007) reported the design and construction of an optical cell allowing direct observation of the mixing in channel tees using an appropriate dye. They observed that convective flow can strongly influence the observed mixing patterns for some configurations of the tees, but cautioned that only the laminar flow regime was explored and their results are, therefore, not directly applicable to most of the common hydrothermal synthesis apparatuses. Middelkoop et al. (2009) reported the use of X-ray measurement techniques (a combination of tomography angle and tomography energy-dispersive diffraction imaging, TADDI and TEDDI, respectively, which they termed TÆDDI) for probing directly the crystallization environment inside a stainless steel mixing unit. Using TÆDDI, they were able to observe the steady state formation of $CeO_2$ nanoparticles, the location of particle growth together with indications of crystallite sizes, and the progressive build-up of material on the mixer/reactor walls during long operation. The focus of the study, however, was on the demonstration of the capabilities of TÆDDI rather than on mixing optimization.

Based on the insights gained through the various approaches, several mixing geometries have been proposed, implemented and tested (see Figure 2.9 for some examples), leading to efficient mixing units for the near-critical and supercritical water reaction

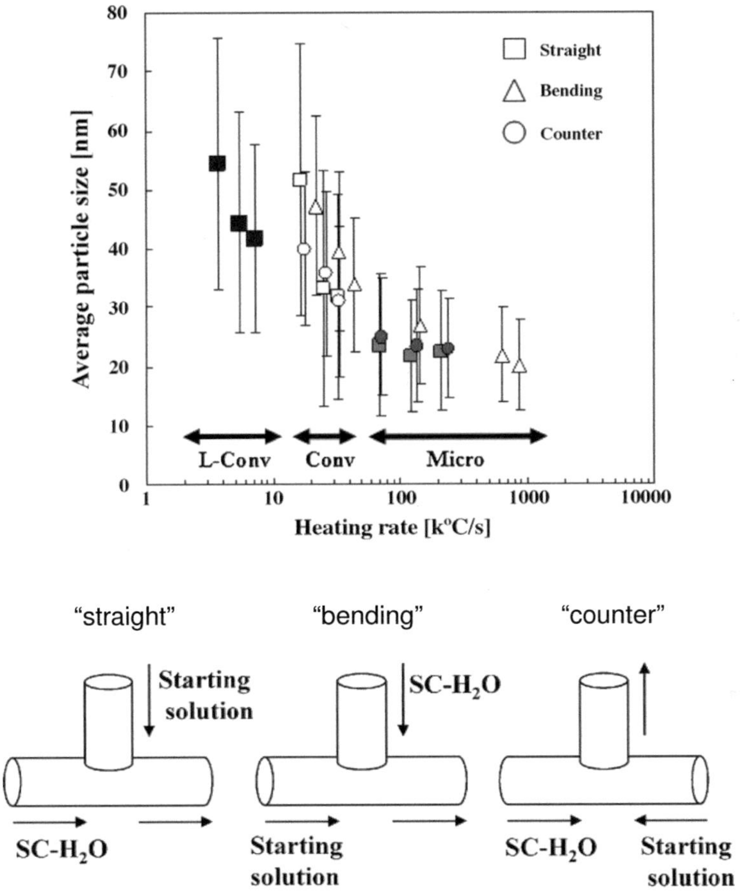


**Figure 2.8:** Relationship between average particle size of hydrothermally synthesized NiO nanoparticles (determined on the basis of TEM images) and the heating rate for three different T-shaped mixers: "L-Conv" with I.D. = 2.3 mm, "Conv" with I.D. = 1.3 mm, and "Micro" with I.D. = 0.3 mm, and three different mixing setups shown at the bottom of the Figure. Figures were taken from (Kawasaki et al., 2010) and used with permission from Elsevier.

and solvent environments. Nevertheless, control of the mixing remains to date one of the key challenges in the development of continuous hydrothermal processes for the synthesis of metal oxide nanoparticles (e.g. Lester et al., 2010).

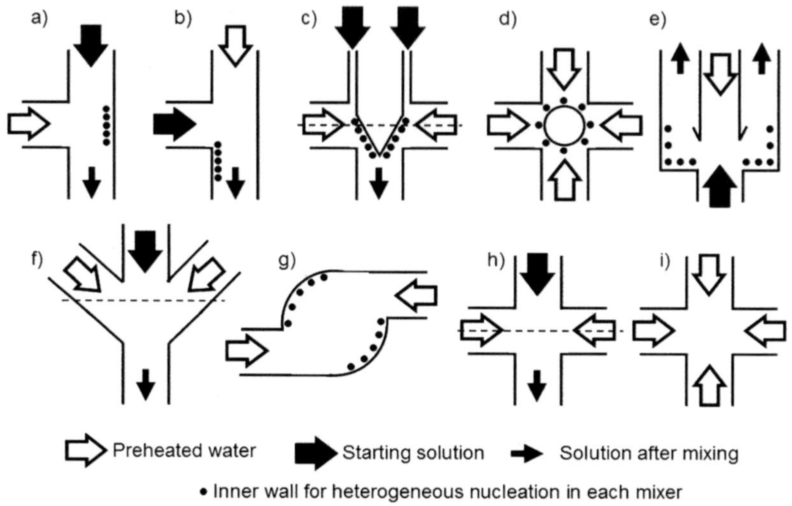


**Figure 2.9:** Mixing geometries developed and applied for the continuous hydrothermal synthesis of metal oxide nanoparticles: a) and b) vertical cross sections of conventional T-type mixers, c) and d) vertical and horizontal cross section of a mixer with a movable needle, e) vertical cross section of a nozzle-type mixer, f) and g) vertical and horizontal cross section of a swirl-type mixer, h) and i) vertical and horizontal cross section of a central collision-type mixer. Figure taken from Sue et al. (2010) and used with permission from the American Chemical Society.

### 2.1.3 In situ functionalization of metal oxide nanoparticles

Due to the special properties of near-critical and supercritical water (see Section 2.1.1), it is possible to add substantial amounts of organic molecules such as thiols, alcohols, aldehydes, carboxylic acids or amines as organic modifiers to form homogeneous nc-$H_2O$ and/or sc-$H_2O$ reaction mixtures (Mousavand et al., 2007; Rangappa et al., 2008; Ziegler et al., 2001). This opens the possibility to synthesize nanocrystals under in situ functionalization with organic capping molecules, providing a further control of the particle formation process with respect to their size and morphology as well as with respect to the surface chemistry of the particles. Both batch and continuous processes for the hydrothermal synthesis of metal oxide/metal nanoparticles can thus be modified to produce organic-inorganic hybrid particles by in situ functionalization with organic molecules. Particles synthesized with a stable surface modifier layer often show narrow size distributions and can be transferred into nonpolar organic solvents where they frequently exhibit a very good dispersibility (Arita et al., 2010). Some illustrative examples from both batch and continuous hydrothermal syntheses are described below. A general scheme for in situ functionalization is shown in Figure 2.10.

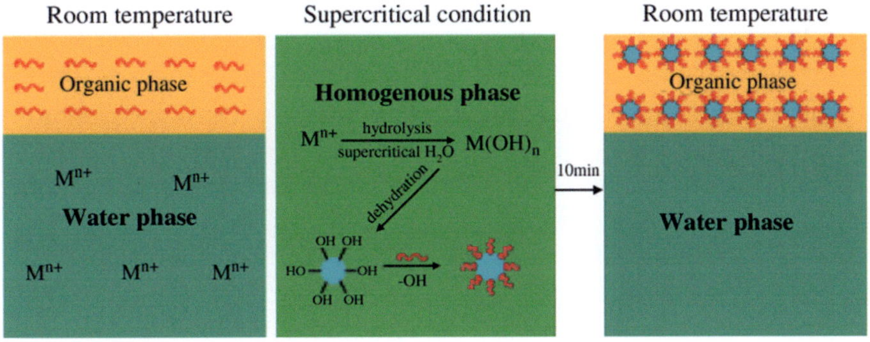


**Figure 2.10:** Scheme of the in situ functionalization of metal oxide nanoparticles with organic ligands. The graphic is taken from Byrappa et al. (2008) and used with permission from Elsevier.

The first known report for in situ functionalization of metal oxide nanoparticles in a hydrothermal process is that of Ziegler et al. (2001). Using a semi-batch process, they reported the use of hydrocarbon capping ligands to stablize nanocrystal formation in supercritical water, yielding smaller particles (by a factor of two comparing capped vs. non-capped particles) and enabling their transfer to and redispersion in an organic solvent (i.e. chloroform). Furthermore, the presence of the ligand can alter the composition of the nanoparticles obtained: Hydrothermal synthesis led to the formation of either copper oxide nanoparticles or copper metal nanocrystals in the absence and presence of the organic ligand (1-hexanethiol), respectively. Finally, altering the reaction conditions (e.g. precursor, pH, capping ligand) significantly affected the morphology, size and degree of agglomeration of the resulting nanoparticles.

Subsequently, several in situ functionalized metal oxide nanoparticles have been synthesized mainly in batch mode. Here, the metal precursor solutions are filled into pressure-resistant tubes (usually 5 mL of inner volume , the inner diameter is generally not reported) to which the organic molecules (mainly C6 to C12 hydrocarbons with one functional group such as an aldehyde, primary amine, carboxyl or hydroxyl) closed tightly and put into an electric furnace at a constant temperature for a certain period of time (mostly 10 min) after which the tubes are removed from the furnace and submerged in a water bath of room temperature for rapid cooling (e.g. Mousavand et al., 2006; Sahraneshin et al., 2012b).

Using this approach, the synthesis and morphology control of e.g. surface functionalized $Fe_2O_3$, AlOOH, $CeO_2$, $Co_3O_4$, nanoscale yttrium aluminium garnet particles (YAG), hafnium oxide ($HfO_2$) has been demonstrated (Arita et al., 2010; Mousavand et al., 2007, 2006; Sahraneshin et al., 2012a,b). The optimum conditions for the surface modification differ among the modifiers and metal oxides. Often, the presence of the organic capping agent does not alter the crystal structure, but greatly alters the surface morphology and surface charge of nanoparticles, by controlling the nucleation and crystal growth. Furthermore, the size of the particles can be controlled by varying the molar ratio of the modifier to starting material (Byrappa et al., 2008). As an example, TEM images of cobalt oxide nanoparticles synthesized with and without surface modifiers are shown in Figure 2.11. The pronounced effect of the organic ligand

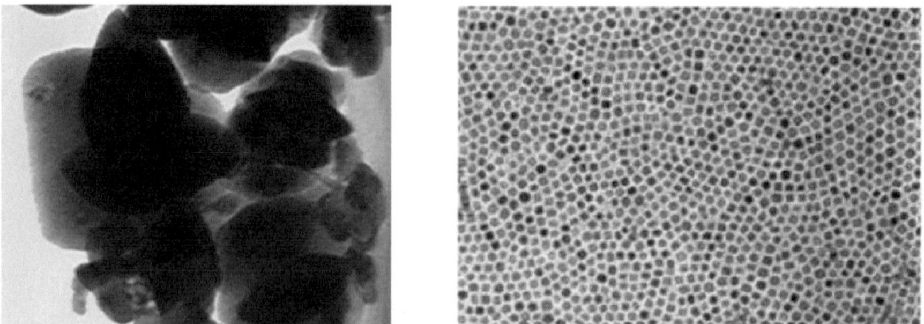


**Figure 2.11:** TEM images of $Co_3O_4$ nanoparticles hydrothermally synthesized at 573 K and 20 MPa: (left) without surface modifier; (right) with surface modifier decanoic acid. Reported in and taken from (Byrappa et al., 2008) and used with permission from Elsevier.

molecules on the morphology and size of $CeO_2$ nanoparticles is shown schematically in Figure 2.12.

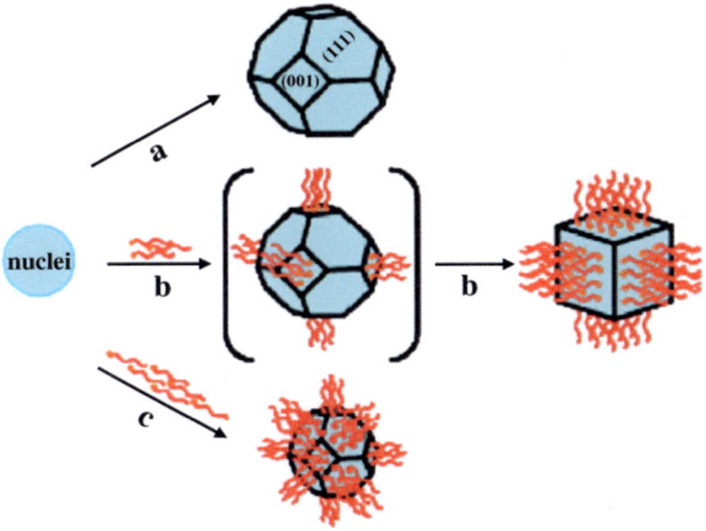


**Figure 2.12:** Schematic on the shape control in the hydrothermal synthesis of ceria nanoparticles: a) without functionalization: truncated octahedrons; b) at a low ratio of decanoic acid to ceria: preferential interaction of the ligand molecules with the ceria (001) planes slows the growth of (001) faces relative to (111) faces leading to the formation of nanocubes; c) at a high ratio of deconoic acid to ceria: organic ligand molecules block growth on both (001) and (111) faces leading to the formation of truncated octahedral and smaller crystals. Figure taken from Byrappa et al. (2008) and used with permission from Elsevier.

In addition to organic molecules bearing one functional group, also bifunctional molecules have been used as modifiers (Taguchi et al., 2012; Takami et al., 2008; Togashi et al., 2011). For example, monoclinic $ZrO_2$ nanoparticles have been synthesized in a batch-type reator in the presence of various carboxylic acids with or without a second functional group at the opposite end (see Figure 2.13a). Here, as in most of the studies on in situ functionalized metal oxide nanoparticles, the surface modification was investiagted by means of FTIR spectroscopy. From the FTIR spectra (see Figure 2.13b) it was concluded that the modifiers always attachted to the surface of the nanoparticles through coordination bonds between the carboxylate group (-$COO^-$) and the

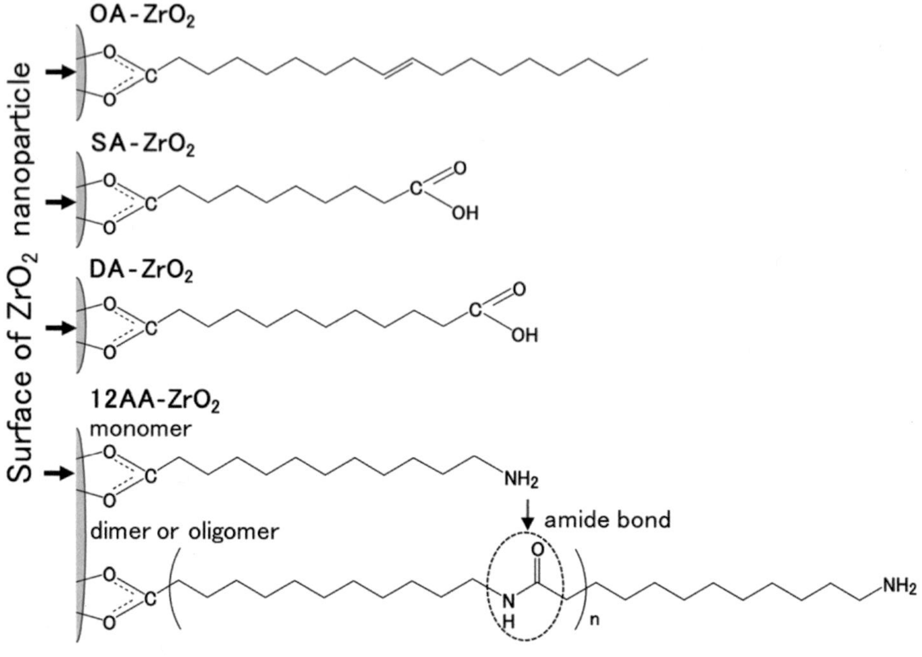
Surface of ZrO2 nanoparticle
OA-ZrO2
SA-ZrO2
OH
DA-ZrO2
OH
12AA-ZrO2
monomer
NH2
amide bond
dimer or oligomer
NH2

(a)

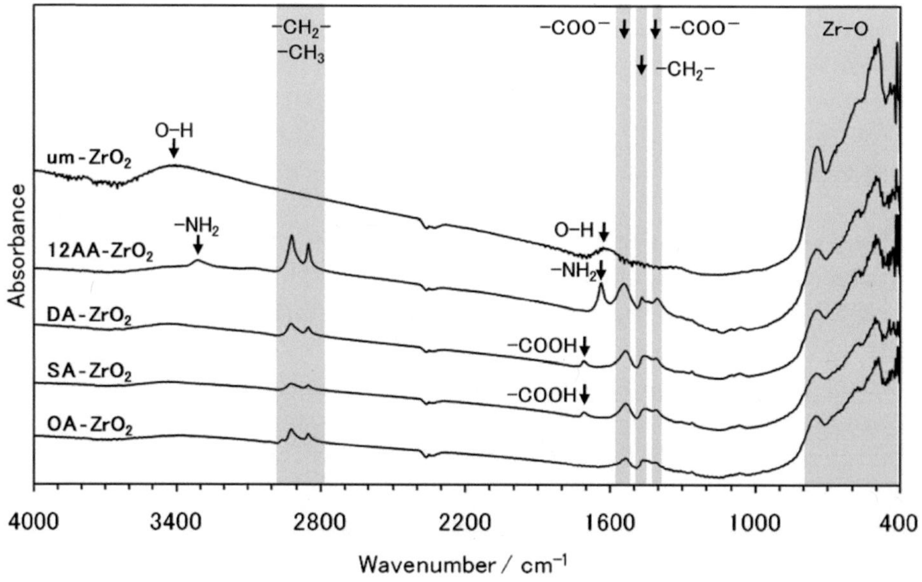
-CH2-
-CH3
-COO−
-COO−
-CH2-
Zr-O
O-H
um-ZrO2
-NH2
12AA-ZrO2
O-H
-NH2
DA-ZrO2
-COOH
SA-ZrO2
-COOH
OA-ZrO2
Absorbance
4000
3400
2800
2200
1600
1000
400
Wavenumber / cm−1

(b)

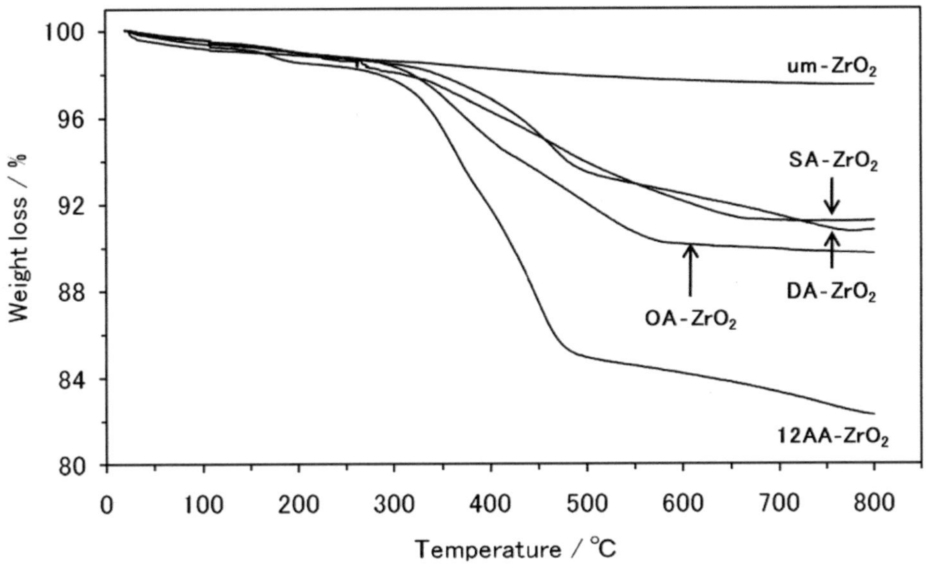


(c)

**Figure 2.13:** Synthesis of surface-modified monoclinic $ZrO_2$ nanoparticles using different carboxylic acids as surface modifiers: with oleic acid (OA- $ZrO_2$), sebacic acid (SA- $ZrO_2$), dodecanedioic acid (DA- $ZrO_2$), and 12-aminododecanoic acid (12AA- $ZrO_2$). a) Schema of the modified surfaces, b) FTIR spectra of the surface modified particles shown in a) and of unmodified $ZrO_2$ (um-$ZrO_2$), and c) TGA curves of the same particles: the number of the surface modifiers attached to the surface was about the same among OA-, SA-, and DA-$ZrO_2$. However, the number of attached 12AA in 12AA-$ZrO_2$ was four to eight times higher than the corresponding organic molecules attached to the other surface modified nanoparticles. The higher amount of attached 12AA was attributed to the formation of dimer and/or oligomer on the surface of 12AA-ZrO2 as shown in a). Figures are taken from Taguchi et al. (2012) and reproduced by permission of The Royal Society of Chemistry.

Zr ion. Furthermore, the presence of the other functional groups, amine (-$NH_2$) and carboxyl (-COOH) was also revealed in these spectra. To affirm the coordination of the modifier to the metal oxide surface, thermogravimetric analysis was carried out, which also provided an estimate for the amount of attachted surface modifiers. The nanoparticles presenting carboxyl and amine functional groups were dispersible in water.

Nanoparticle synthesis in the presence of bifunctional ligands has also been carried out in a continuous process: Takami et al. (2008) reported on the synthesis of $CeO_2$ nanocrystals in the presence of dicarboxylic acids ($HOOC(CH_2)_xCOOH$). They prepared aqueous $Ce(NO_3)_3$ solutions containing hexanedioic acid and heated them rapidly in a typical hydrothermal lab-scale plug-flow reactor ($T$ = 523 and 573 K, $p$ = 25 MPa) by mixing with a hot compressed water stream (cf. Figure 2.4). With this setup the dicarboxylic acid can serve as an interparticle linker, leading to the formation of cubic nanoassemblies of octahedral $CeO_2$ primary nanoparticles (see Figure 2.14).

Simlarly, Lu et al. (2012) synthesized in situ functionalized tin-doped indium oxide (ITO) nanoparticles by using a continuous hydrothermal flow reactor ($T$ = 723 K ; $p$ = 30 MPa, residence time $\approx$ 7 s), in which the ligands and precursor solutions both pass through the reaction section. The use of the organic modifier (hexanoic acid dissolved in hexane) led to highly crystalline, hydrophobic nanoparticles. Also using a continuous setup, Xu and Teja (2008) synthesized iron oxide in the presence of a polymer (polyvinyl alcohol, PVA) dissolved in the precursor solution. They found that addition of PVA led to narrow particle size distributions and that the concentration of PVA also influenced particle size. TEM images of the resulting particles are shown in Figure 2.15. Another example for the use of polymers as modifiers during particle formation is the synthesis of $CeO_2$ in the presence of either PVA or polyacrylic acid (PAA), as reported by Taguchi et al. (2011) using a batch reactor.

Organic modifiers have also been introduced immediately after the reaction section of continuous hydrothermal plants by adding the modifiers (presumably as aqueous solutions) before the cooler: Hayashi et al. (2010b) reported on the synthesis of $BaTiO_3$ from $TiO_2$ sol/$Ba(NO_3)_2$ mixed solutions at $T$ = 673 K, $p$ = 30 MPa and $\tau$ = 8 ms. They found that the primary particle size did not change when using different dispersants (e.g. PAA, Tween 80) but the identity of the dispersant had an effect on the agglomeration behavior of the particles in the resulting aqueous suspensions.

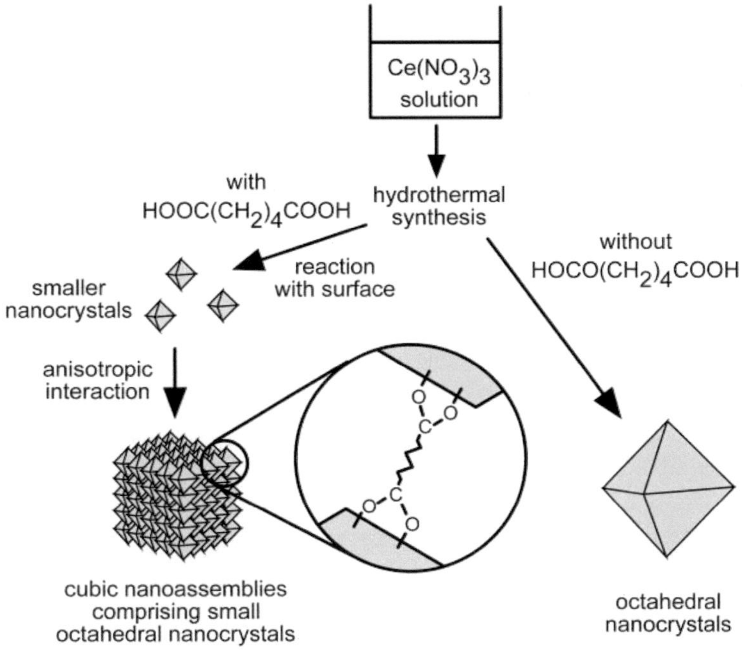


**Figure 2.14:** Proposed mechanism for the formation of cubic nanoassemblies from octahedral $CeO_2$ nanocrystals synthesized in the presence of hexanedioic acid ($HOOC(CH_2)_4COOH$) using a plug-flow reactor. The two carboxyl groups of hexanedioic acid are bound to the Ce atoms on the surface of the cerium oxide nanocrystals and hereby preventing the growth of the primary octahedral $CeO_2$ nanocrystals and assemble the primary nanocrystals to form superstructures. Figure taken from Takami et al. (2008) and reproduced by permission of The Royal Society of Chemistry.

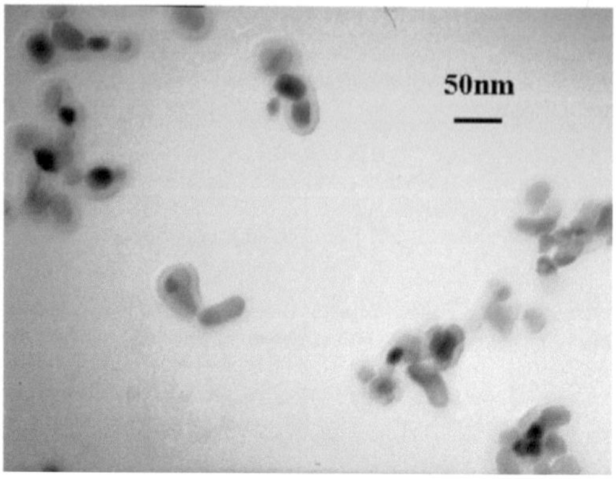


**Figure 2.15:** TEM image of iron oxide nanoparticles synthesized in the presence of polyvinyl alcohol (PVA) in a continuous hydrothermal process at 573 K (mixing point) and 21.7 MPa. Figure taken from Xu and Teja (2008) and used with permission from Elsevier.

## 2.2 Iron oxide nanoparticles

### 2.2.1 Properties and applications of iron oxide (nanoparticles)

Iron oxide (nano)particles play an important role in natural processes and have manifold technical applications based on their special and diverse physical, chemical, and structural properties. Applications of iron oxide nanoparticles are found in drinking water and waste water treatment, electronics, optoelectronics, diagnostics, therapeutics, and bioseparation. There is a long tradition in the (interdisciplinary) investigation of iron oxides because of their prevalence in nature and their use in technical applications. Although there is a large amount of information available on the 16 identified iron oxides[†], there are still open questions especially regarding the size-dependent reactivity and chemical structure of iron oxide nanoparticles (Chernyshova et al., 2007; Cornell and Schwertmann, 2003).

For the work described in this thesis, the relevant iron oxides are hematite, magnetite, maghemite and ferrihydrite, of which selected properties are shown in Table 2.1. The first three have been used as pigments for thousands of years. Furthermore, hematite

[†]This term includes iron oxides, iron hydroxides and iron oxide-hydroxides where iron can take the oxidation states $Fe^{3+}$ and/or $Fe^{2+}$.

and maghemite have been used as catalysts for a long time; magnetite and maghemite are used on account of their magnetic properties especially in electromagnetic storage media (Cornell and Schwertmann, 2003). Furthermore, nanometer sized maghemite or magnetite particles exhibiting superparamagnetic behavior (see below) dispersed in aqueous or organic media - so called ferrofluids - have proven to be useful in a broad range of applications besides its original design intention as high-performance seals in space applications. Due to their special behavior which is mostly based on the fact that the ferrofluids only have a net magnetic moment upon applying a magnetic field, they are currently employed e.g. as seals in vibrating and rotating environments such as computer discs (Teja and Koh, 2009). In addition, these iron oxides are considered biocompatible and show low toxicity, which has led to significant research into their use (as nanoparticles) in diagnostics and medical technology (Teja and Koh, 2009). Superparamagnetic magnetite nanoparticles have been used in vivo as magnetic resonance imaging (MRI) contrast agents for molecular and cell biolgy. Furthermore, they are in use as hyperthermia inducing agents in cancer therapy (Hayashi et al., 2010a). In general, there is a lot of potential by using these iron oxide nanoparticles with specially designed surface functionalizations simultaneous in cancer cell imaging and treatment (Chen et al., 2011; Yu et al., 2008). Besides these in vivo applications, the magnetic iron oxide nanoparticles have been applied also in cell separation, organic and biochemical syntheses, protein purification, capturing toxic metals from liquid matrices for industrial applications as well as environmental remediation (Rutledge et al., 2010; Teja and Koh, 2009).

Also, ferrihydrite, of which the structure is still under debate (see below) despite its critical role in environmental geochemistry, has been used in industrial applications such as direct coal liquefaction and matallurgical processing (Michel et al., 2007; Peak and Regier, 2012). The very high surface area and functional group density enables ferrihydrite to sequester various species through adsorption, coprecipitation, and redox reactions (Smith et al., 2012). Therefore, it has been manufactured as scavenger of heavy metals and metalloids in the area of waste water treamtent and remedial acitivities. Futhermore, ferrihydrite is suspected to form the inorganic core of ferritin, an iron storage protein which plays a key role in controlling the iron level in plants, animals, and microbes (Michel et al., 2007).

A number of investigations on iron oxide nanoparticles have shown that their structure and properties strongly depend on the particle size, on their history of formation and on the medium surrounding them, which makes their characterization considerably more difficult (Chernyshova et al., 2007; Cornell and Schwertmann, 2003). In the following, some relevant aspects in the context of this work on the size-dependant structural transformations and properties of the above mentioned iron oxides are introduced.

**Table 2.1:** Some properties of the iron oxides relevant to this work adapted from Cornell and Schwertmann (2003) and Teja and Koh (2009) with additions for ferrihydrite from Tüysüz et al. (2008).

| Mineral name | hematite | magnetite | maghemite | ferrihydrite |
|---|---|---|---|---|
| Crystal system | rhombohedral hexagonal | cubic | cubic or tetragonal | (hexagonal) |
| Structural type | Corundum | Inverse spinel | Defect spinel | |
| Chemical formula | $\alpha$-$Fe_2O_3$ | $Fe_3O_4$ | $\gamma$-$Fe_2O_3$ | ($Fe_5HO_8$ $\cdot 4H_2O$) |
| Density ($g/cm^3$) | 5.26 | 5.18 | 4.87 | 3.96 |
| Type of magnetism under standard conditions | weakly ferromagnetic | ferrimagnetic, superparamagnetic in the nano-regime | ferrimagnetic, superparamagnetic in the nano-regime | superparamagnetic |
| Saturation magnetization at 300 K ($Am^2/kg$) | 0.3 | 92-100 | 60-80 | |
| Color | red | black | reddish-brown | red-brown |

## Color

The broad range of colors exhibited by iron oxides demonstrates in the most obvious way the transition character of iron and the complex crystals and electronic structures of its compounds leading to continous or sudden changes in their chemical and physical properties. The color can be used as a starting point to dinstinguish between the different iron oxides. However, the degree of crystallinity, the particle size and shape, degree and form of aggregation and cation substitution cause variations in the color of a particular iron oxide. For example and important for the work discribed in this thesis, suspensions of hematite particles are known to change their color from orange to red to even purple upon increasing particle sizes. Also strong changes in color may result from oriented aggregates of nanocrystallites, as has been reported for hematite particles (Chernyshova et al., 2007; Cornell and Schwertmann, 2003).

### Magnetic properties of iron oxide (nanoparticles)

Iron atoms possess a strong magnetic moment arising from four unpaired electrons in their 3*d* orbital. When crystals are formed containing iron, different types of magnetic behavior may result (the following is based on: Cornell and Schwertmann, 2003; Riedel, 1990; Schultz, 2007; Teja and Koh, 2009):

- *Paramagnetism* is characterized by a zero net magnetic moment as the result of randomly oriented, magnetically isolated atomic magnetic moments. However, these moments may become partly oriented under the influence of an external magnetic field, leading to a small non-zero net magnetic moment. This magnetic moment disappears when the external field is removed.
- In contrast to paramagnetic materials, the individual magnetic moments of *ferromagnetic* materials remain aligned once the external magnetic field has been removed. Moreover, the strength of the magnetic fields from ferromagnetic materials is much larger than that of paramagnetic materials.
- *Ferrimagnetism* arises when two types of atoms within the crystal give rise to two net magnetic moments aligned in an antiparallel fashion.
- If the two antiparallel net magnetic moments in ferrimagnetic crystals are of the same magnitude, the net magnetic moment is zero and the crystals are said to be *antiferrimagnetic*.
- *Superparamagnetism* arises in crystals containing only one magnetic domain (i.e. crystal sizes in the nano-regime) as the result of the existence of crystallographic directions along which the electron spins are preferentially aligned (magnetic anisotropy). The direction of magnetization within this single magnetic domain can change rapidly between the preferential directions if enough thermal energy is available to overcome the energy barrier for this transition. This is often the case for iron oxide nanoparticles $< 20$ nm at room temperature. The result of these changes in magnetization direction is a reduced or absent net magnetic field. These rapid transitions can be stopped by the introduction of an external magnetic field of sufficient strength, leading to a response like paramagnetic materials but with a strength reminiscent of ferromagnetic materials.

In addition to the type of the magnetism which arises from the structure of the specific iron oxide, particle size, morphology, degree of crystallinity, and surface effects (e.g. surface functionalization) can strongly influence the magnetic properties of the iron oxide nanoparticles. In consequence, the magnetic properties of iron oxide nanoparticles can be tuned by the directed nanostructuring (the preparation method including surface treaments) of the iron oxide (Teja and Koh, 2009).

#### Structure of hematite in the nano-regime

Chernyshova et al. (2007) undertook a rigorous study on the structural transformation of hematite with decreasing particle size with the aim of addressing unresolved questions related to the structure of hematite particles in the nano-regime, some dating back more than 30 years. Within this study, it was found that hematite particles are destabilized in the nano-regime by possessing maghemite-like defects in the near-surface region. It was concluded that the formation of the structural defects is governed by the thermodynamics of the $\alpha \rightarrow \gamma$ phase transition with decreasing particle size, driven by the need to compensate for the increased contribution of the particle's surface on the free energy of the particles. The onset of the phase transition was found to be highly dependant on the environment and on the growth kinetics of the particles. Importantly, the structure of hematite nanoparticles is significantly stabilized in a water matrix, which was attributed to the surface energy loss provided by the electrochemical double layer effect and the increased rigidity of the particle surrounding. For the dry particles used in the study of Chernyshova et al. (2007), the onset of the $\alpha \rightarrow \gamma$ phase transition was observed at particle diameters of about 40 nm. As a consequence, the particles need to be characterized in the medium in which they were synthesized or are to be used to prevent artefacts due to possible structural changes upon changing the environment of the particles e.g. upon drying.

#### Ferrihydrite

Ferrihydrite is metastable in natural systems and occurs as discrete particles as well as surface coatings in soils and sediments. It is formed as initial solid produt of Fe(III) hydrolysis in solution and is easly transformed into more crystalline iron oxide forms (Peak and Regier, 2012). The structure of ferrihydrite is still a matter of debate despite considerable efforts. Reasons for this are the difficulty in probing ferrihydrite because it only exists in the nanoparticulate form (individual crystallites $< 10$ nm) and these crystallites are succeptible to structural changes upon changes in their environment. In addition, no single formula is widely accepted for ferrihydrite, which is often described as "two-line" or "six-line" ferrihydrite depending on the number of broad, poorly-defined maxima to be observed in PXRD patterns (Michel et al., 2007). This distinction, however, arises from differences in the size of their coherent scattering domains or crystallite sizes rather than on significant differences in structure, which have been shown to be very similar (Smith et al., 2012).

### 2.2.2 Synthesis of iron oxide nanoparticles using near- and supercritical water

The synthesis of hematite, magnetite, maghemite, and ferrihydrite in near- and/or supercritical water has been reported (Lam et al., 2008; Sue et al., 2010). For this, batch-type as well as continuous processes have been applied. Concerning the former type of setup, the formation of hematite nanoparticles could also take place at the porous surface of added activated carbon pellets resulting in an activated carbon/iron oxide nanocomposite applicable as a catalyst e.g. in the oxidation of propanal (Teja and Koh, 2009). Concerning the continuous variant, two types of setups have been employed. In case of the method developed from Matson and co-workers, known as rapid thermal decomposition of precursors in solution (RTDS), the iron salt solution is directly brought up to supercritical conditions via external heating to the target temperature. Particle nucleation and growth are terminated by rapid cooling (achieved by spraying the solution into a flask submerged in an ice-bath; Matson et al., 1992). The other type of continuous setup is the one developed by Adschiri and Arai which was described in detail in Section 2.1.2 and is applied in this work. Different iron salt solutions as starting material have been used in both variants as well as various reaction times and operating conditions (temperature and pressure) leading to different iron oxide modifications, particle sizes and morphologies.

Using a setup of this latter type, Lam et al. (2008) reviewed the synthesis of iron oxide nanoparticles from iron nitrate, iron sulfate and iron chloride solutions. The particles ranged in size from five to 50 nm (as measured from TEM images) and the structure was interpreted to be that of hematite. Xu and Teja (2008) found that the morphology of the majority of the synthesized particles changed from spherical to rhombic with increasing educt concentration when synthesizing hematite particles from iron nitrate solutions. In the same study, they observed that the particle size did not change significantly when varying the temperature at the mixing point in the range 487–648 K at a pressure of 23 MPa.

The synthesis of magnetite nanoparticles has been carried out from aqueous iron ammonium citrate solutions (pressumably through the reduction of $Fe^{3+}$ ions by carbon monoxide produced by the decomposition of citrate under hydrothermal conditions) and from iron(II) acetate (Cabanas and Poliakoff, 2001; Lam et al., 2008). Crystallites smaller than 10 nm are produced from aqueous iron ammonium citrate solutions (Bremholm et al., 2008) while biomodal distributions in the range 10–100 nm are found in particles synthesized from iron(II) acetate solutions (Cabanas and Poliakoff, 2001). However, it is worth noting that the fluid velocities used in this study are low compared to the more recent publications.

In fairly recent studies, the synthesis of iron oxide from aqueous iron(III) nitrate solutions was selected as a model reaction for evaluting different types of mixers. For example, Sue et al. (2010) compared a central collision-type micromixer and a T-type micromixer (inner diamter = 0.3 mm) at 673 K and 30 MPa (see Figure 2.9 for mixer configurations). They examined the effect of varying the iron content ($c_{\mathrm{Fe,\,after\,mix}}$ = 0.01, 0.05, 0.1 mol $kg^{-1}$) and residence time (0.1–1 s) on the average particle size and particle size distribution based on TEM images and PXRD, and placed these results in the context of the estimated $Fe_2O_3$ solubility (which decreases with increasing temperature and decreasing $Fe(NO_3)_3$ molality) and corresponding supersaturation under hydrothermal conditions (see Figure 2.16).

In all experiments performed by Sue et al. (2010), $Fe^{3+}$ conversion was more than 96% and particles in the size range 4–80 nm (with average particle sizes less than 16 nm) were observed in TEM images. Both the average particle size and the corresponding coefficient of variation increased with increasing $Fe(NO_3)_3$ molality, irrespective of the type of micromixer. Whereas at low iron content and short residence times very similar particle size distributions were obtained for both types of mixers, the coefficient of variation was markedly decreased when using the central-collision-type micromixer compared to the T-type micromixer at higher iron content and longer residence times (see Figure 2.17). Especially at 0.50 mol $kg^{-1}$ and 1 s, the particle size distribution changed from bimodal to monomodal by using the central-collision-type vs. the T-type micromixer. Furthermore, the structure of the crystallites was examined by PXRD after drying of the particles. At low $Fe(NO_3)_3$ concentration, the product was found to be $\gamma$-$Fe_2O_3$ at 0.1 s for both types of micromixer. With increasing residence time to 1.0 s, part of the product changed to $\alpha$-$Fe_2O_3$. At high $Fe(NO_3)_3$ molality, the product obtained using the T-type mixer was a mixture of $\gamma$-$Fe_2O_3$ and $\alpha$-$Fe_2O_3$ at 0.1 s but $\alpha$-$Fe_2O_3$ became dominant at 1.0 s. In contrast, the product obtained using central-collision-type mixer was $\gamma$-$Fe_2O_3$ (single phase) at 0.1 s and varied to a mixture of $\gamma$-$Fe_2O_3$ and $\alpha$-$Fe_2O_3$ at 1.0 s.

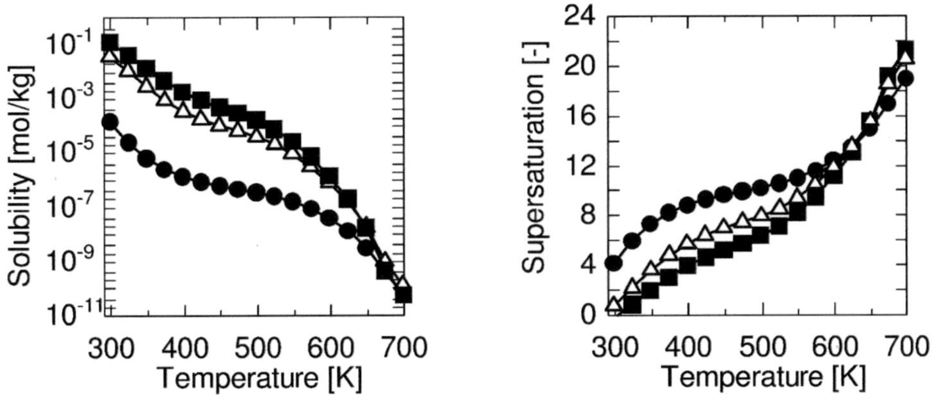


**Figure 2.16:** Estimated solubility (left) and corresponding supersaturation (right) of $Fe_2O_3$ in water at 30 MPa and 0.01 mol $kg^{-1}$ (circle), 0.05 mol $kg^{-1}$ (triangle) and 0.10 mol $kg^{-1}$ (square) of $Fe(NO_3)_3$. The degree of supersaturation was estimated as the logarithm of the ratio of the $Fe(NO_3)_3$ molality after mixing to the calculated solubility of $Fe_2O_3$. An increase in $Fe(NO_3)_3$ molality results in an increase in $HNO_3$ molality and also in $Fe_2O_3$ solubility. At temperatures below $T_c$, $HNO_3$ dissociates completely and an increase in $Fe(NO_3)_3$ molality by a factor of 10 results in an increase in $Fe_2O_3$ solubility by more than 1000 times. In contrast, at temperatures above $T_c$, $HNO_3$ strongly associates because of the low permittivity. Therefore, the $Fe_2O_3$ solubility changes only slightly (less than a factor of two) when the molality of $Fe(NO_3)_3$ is increased by one order of magnitude. Figure taken from Sue et al. (2010) and used with permission from the American Chemical Society.

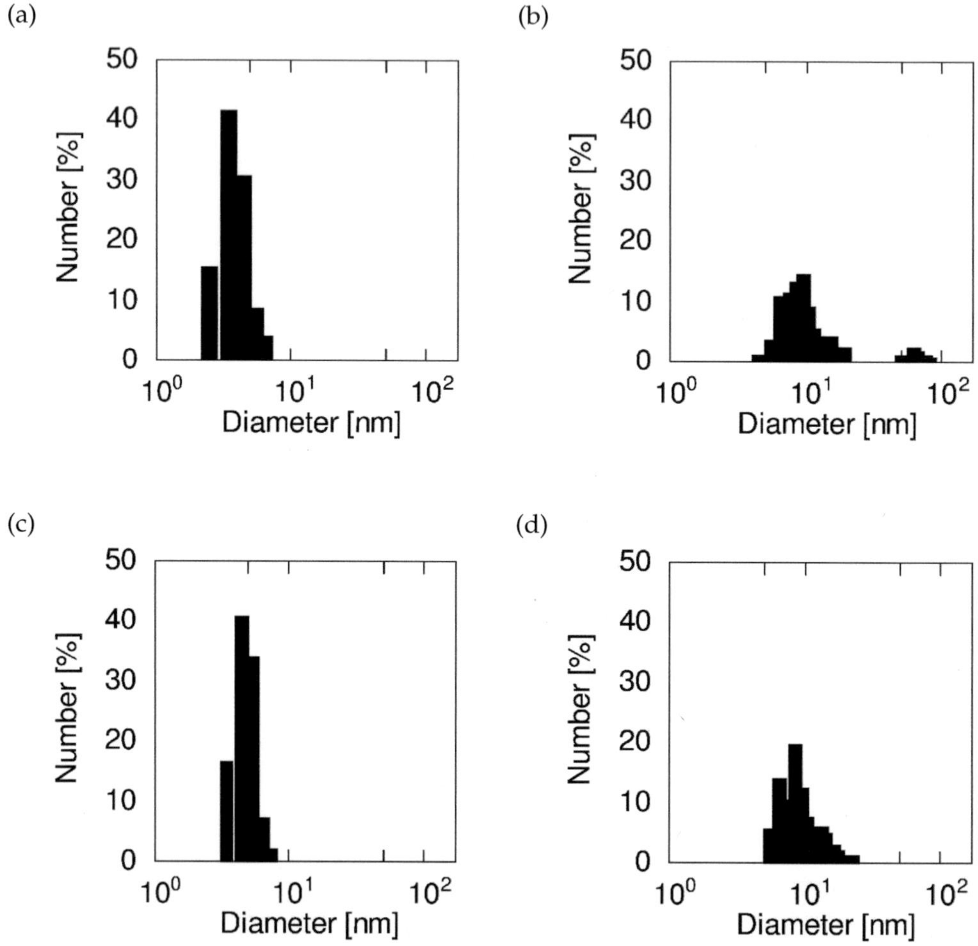


**Figure 2.17:** Particle size distributions determined on the basis of $\approx 200$ particles measured from TEM images. Particles were synthesized from iron(III) nitrate solutions using a T-type micromixer at (a) $c_{\mathrm{Fe,\,after\,mix}} = 0.01$ mol kg$^{-1}$ and $\tau = 0.1$ s, (b) $c_{\mathrm{Fe,\,after\,mix}} = 0.1$ mol kg$^{-1}$ and $\tau = 1$ s, and using a central collision-type micromixer (see Section 2.1.3, Figure 2.9, (c) $c_{\mathrm{Fe,\,after\,mix}} = 0.01$ mol kg$^{-1}$ and $\tau = 0.1$ s, and (d) $c_{\mathrm{Fe,\,after\,mix}} = 0.1$ mol kg$^{-1}$ and $\tau = 1$ s. Diagrams are taken from Sue et al. (2010) and used with permission from the American Chemical Society.

Studies where the synthesized particles are functionalized in situ are of special interest because of the additional possibilities they open. For example, the particle size and the breadth of the particle size distributions can be significantly reduced through in situ functionalization with hexylamine, hexanoic acid (Adschiri et al., 2006, 2005) and poly(vinyl alcohol) (Xu and Teja, 2008). These and other possibilities of in situ functionalization are described in Section 2.1.3. In addition to the examples, described there, Togashi et al. (2011) reported the one-pot synthesis of in situ functionalized $Fe_3O_4$ nanoparticle clusters, in which the nanoparticles were interlinked by 3,4-dihydroxycinnamic acid. The clusters ranged in size from 50 to 400 nm and were composed of nanoparticles of approximately 20 nm in diameter. The average size of the clusters could be tuned within this range by changing the reaction time. On account of the large number of conjugated COOH groups, these clusters could be dispersed in water up to concentrations around 20 g $L^{-1}$. In addition, the clusters could be manipulated by an external magnetic field because of large size and higher magnetization per cluster compared to the single nanoparticles. At the same time, the magnetic response could be tuned by controlling the size of the clusters without affecting the superparamagnetic–ferromagnetic transition, which depends on the size of the individual nanoparticles. These results are especially relevant for biomedical applications because superparamagnetism enables the clusters to be dispersed in water while allowing manipulation by an external magnetic field. Furthermore, the modification of the particle surface with COOH groups allows the conjugation of biomolecules such as proteins, DNA and pharmaceuticals.

## 2.3 Glyconanoparticles

The surface of mammalian cells is covered with a dense coat of complex oligosaccharides, the so-called glycocalyx . Glyconanoparticles—nanoparticles with carbohydrates as functional ligands—are meant to mimic the presentation of the carbohydrate epitopes of these cell surfaces. The oligosaccharides comprising the glycocalyx are covalently bound to lipids and proteins and, therefore, develop their numerous biological functions as glycolipids and glycoproteins and are referred to as glycans (de la Fuente and Penades, 2006; Taylor and Drickamer, 2003). The arrangement of oligosaccharides in three-dimensional space is commonly referred to as their conformation. Analogously to the different tertiary structures of proteins with different sequences, glycans with different covalent structures take up different conformations

(Taylor and Drickamer, 2003). These complex carbohydrate structures are involved in a broad range of physiological and pathological processes such as cell communication and trafficking, immune responses and tumor genesis and progression (Sihelníková and Tvaroška, 2007; Wang et al., 2010). Examples are shown in Fig. 2.18.

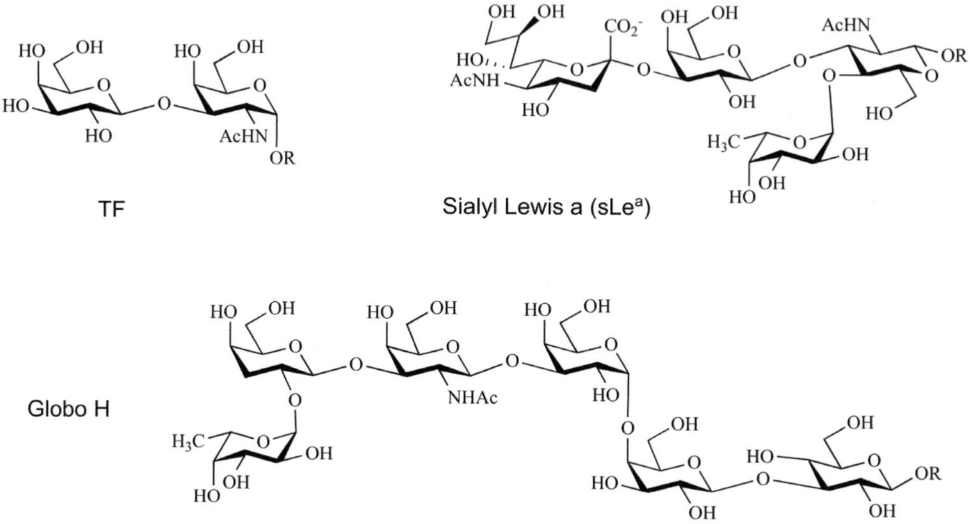


**Figure 2.18:** Terminal glycan epitopes commonly found on tumour tissues (selection from Cipolla et al., 2008) and used with permission from Bentham Science Publishers Ltd.

Hereby, carbohydrate–carbohydrate and carbohydrate–protein interactions are involved (Sihelníková and Tvaroška, 2007). The most important characteristic of these interactions is their extremely low affinity, which is overcome by the multivalent presentation of the ligands and receptors at the cell surface which are clustered to form microdomains ("cluster glycoside effect"). This, as well as the complexity and low abundance of glycans—no simple standardized protocols for their synthesis existed—are reasons why the crucial importance of carbohydrates had been underestimated by the scientific community for a long time (Lepenies et al., 2010; Wang et al., 2010). However, the interest in this field has recently increased dramatically largely due to significant developments in carbohydrate synthesis, glycan analysis methods and nanotechnolgy (Wang et al., 2010). The recent rise in the attention to that field becomes e.g. apparent in the strong increase of published and cited papers on glyconanoparticles (see Figure 2.21).

The combination of the size of the nanoparticles—which is accompanied by a high surface area allowing the accomodation of high-density ligands mimicking the cluster glycoside effect—and their possession of unique and tunable magnetic, optical, electronic, and mechanical properties as well as chemical reactivities allow their incorporation into cells for in vitro and in vivo imaging, drug-delivery, and targeting specific cells such as tumors. The range and potential of possible applications is tremendous and has only started to be explored (Wang et al., 2010).

The first synthesis and scientific application of "glyconanoparticles", reported in 2001, consisted of a gold core and presented the trisaccharides Lewis x ($Le_x$) at their surface. These were used for the study of the self-aggregation of these oligosaccharides (de la Fuente et al., 2001). Thereafter, the synthesis of glyconanoparticles on the basis of gold and silver, quantum dots, or magnetic nanoparticles have been reported (reviews by de la Fuente and Penades (2006); Marradi et al. (2010)). A schematic drawing of the "glyconanoparticle concept" is shown in Figure 2.20. Whereas at the beginning the preparation was mainly driven to carry out basic studies on carbohydrate–carbohydrate and carbohydrate–protein interactions, studies towards their use for in vitro and in vivo applications such as anti-cancer agents (e.g. Cipolla et al., 2008), antiviral agents (e.g. against HIV Martínez-Ávila et al., 2009), vaccines (e.g. anticancer Ojeda et al., 2007), diagnosis (e.g. pre-symptomatic in vivo imaging of brain diseases such as multiple sclerosis (MS), ischemic stroke etc. Serres et al., 2009; van Kasteren et al., 2009), and bacterium detection and decontamination (El-Boubbou et al., 2007) have been increasing. Although it is expected that the carbohydrate based shell of the glyconanoparticles lowers the intrinsic toxicity of metal nanoclusters, the major concerns regarding nanotoxicity need to be addressed for in vivo applications (Garcia et al., 2010).

Besides the importance of the exact design of the core, an important and often critical step in the preparation of glyconanoparticles is the surface coupling chemistry for attaching carbohydrates to the nanoparticles (Wang et al., 2010). Two different strategies have been undertaken: noncovalent (through e.g. electrostatic adsorption) and covalent attachment. However, the latter method is preferred for the preparation of stable glyconanoparticles bearing a certain ligand density in the preferred conformation. Here, the most extensively used coupling chemistry is the thiol- and disulfide-Au chemistry which allows the preparation of gold nanoparticles with well-defined surfaces. In this technique, thiol groups are introduced directly via a linker to the carbohydrates. However, a disadvantage of this coupling strategy is the ligand-lability

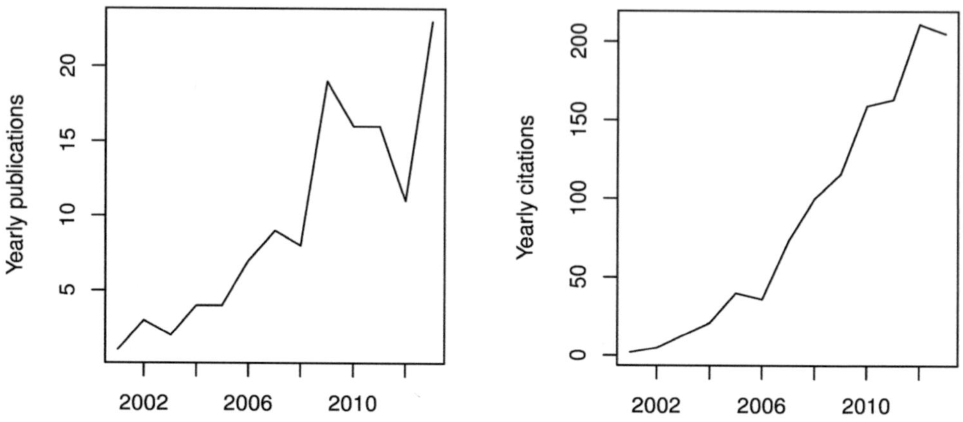


**Figure 2.19:** Number of published and cited papers in the field of glyconanoparticles. Data from www.scopus.com initially accessed November 2012 and updated February 2014; search term: "glyconanopart*".

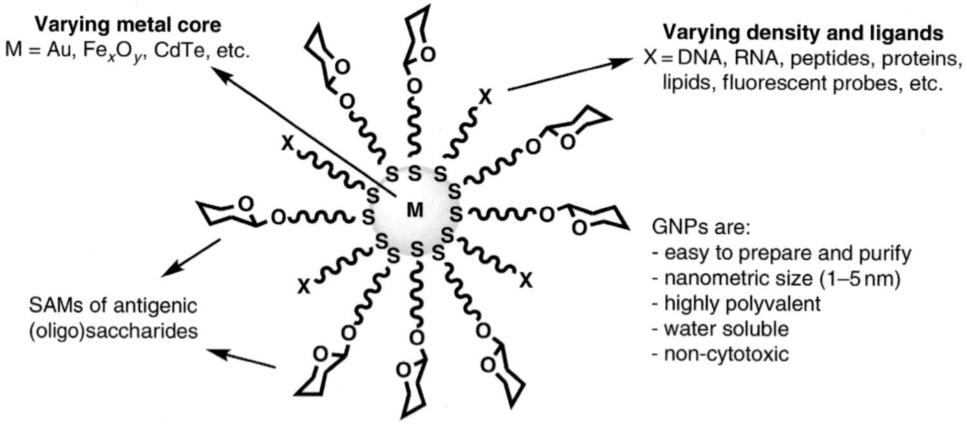


**Figure 2.20:** The glyconanoparticle (GNP) concept: from simple GNPs to complex "nanocells" (Marradi et al., 2010, reproduced with permission from Elsevier).

in biological media. Other coupling methods based on the reaction of complementary functional groups such as the attachment to amine-functionalized dextran coated iron oxide nanoparticles have been reported (van Kasteren et al., 2009). A different approach is the use of un-derivatized carbohydrates. The successful linkage has been reported by a photoinitiated coupling strategy to perfluorophenylazides which were attached to the surface of gold, iron oxide, and silica nanoparticles via common surface chemistry approaches. Another promising approach is the use of *click* chemistry. Using silica coated iron oxide nanoparticles, the attachment of D-mannose via a triazole linker has been reported (El-Boubbou et al., 2007). *Click* chemistry and its application in colloidal science/surface chemistry is discussed in the next section.

## 2.4 *Click* chemistry

The high efficiency exhibited by biological systems has long been a model in the search of synthesis routes with minimal waste. Over the years, concepts such as atom economy (aiming to include every atom from a reagent in the product of the reaction), energy efficiency (aiming to minimize the energy requirements of chemical processes), catalysis (aiming at lower energy input and faster reaction kinetics), step economy (aiming to reach a given product with as few synthetic steps as possible), as well as safety and toxicity issues have been adopted in search of more efficient and environmentally friendly procedures (Lallana et al., 2012).

With these goals in mind and aiming to produce drug candidates, Kolb et al. (2001) defined a set of criteria for selecting reactions belonging to what they termed "*click* chemistry", starting from the observation that carbon-heteroatom connections are used by living organisms to assemble the macromolecules central to life processes together with the conviction that "all searches must be restricted to molecules that are easy to make". They defined *click* reactions as:

- Being modular and wide in scope
- Giving very high yields
- Generating only inoffensive products that can be removed by non-chromatographic methods such as crystallization or distillation
- Being stereospecific but not necessarily enantioselective

- Requiring simple reaction conditions, ideally not being negatively affected by neither oxygen nor water
- Requiring no solvent, or one that is benign, or one that can be easily removed
- Using readily available starting materials and reagents
- Exhibiting simple product isolation
- Yielding a product that is stable under physiological conditions

Surprisingly, there are reactions that fulfill all criteria listed above, a selection of which is shown in Figure 2.21. Common to all of them is a large thermodynamic driving force (usually $> 20\,\mathrm{kcal\,mol^{-1}}$) which drives them to proceed rapidly to completion and usually exhibit good selectivity for a single product (Kolb et al., 2001). With the exception of the copper(I)-catalyzed azide-alkyne cycloaddition (CuAAC, discovered in 2002, see below), most reactions fulfilling the *click* criteria have been known for a long time. For example, the uncatalyzed azide-alkyne reaction is a 1,3 dipolar cycloaddition, a reaction meachnism discovered and systematically explored across a wide range of reactants by R. Huisgen in the 1960s, while the azide-alkyne reaction itself was discovered by A. Michael in 1893. However, by far the most widely used *click* reaction is the CuAAC, which was discovered shortly after the publication of the the *click* chemistry paradigm and has practically become synonymous with it (Adzima and Bowman, 2012).

The CuAAC reaction occurs between an organic azide and a terminal alkyne in the presence of Cu(I) to form a 1,4 disubstituted 1,2,3 triazole. The catalyzed rate of reaction is in the order of $10^7$ times larger than that of the uncatalyzed reaction. The required Cu(I) catalyst may be either added (as a Cu(I) salt such as copper iodide, chloride or acetate) or generated in situ (e.g. by combining Cu(II) salts such as copper(II) sulfate or copper(II) acetate with a mild reductant such as sodium acetate; this approach has the added benefit that the ascorbate also reduces any dioxygen present, thus limiting the formation of any oxidative byproducts). Alternatively, the Cu(I) may be provided by copper metal, resulting in slower reaction kinetics at room temperature but also lower copper contamination levels (Adzima and Bowman, 2012; Such et al., 2012).

Despite its mechanistic complexity (some aspects being still unclear), the CuAAC proceeds highly efficiently in many solvents, including water, is unaffected by most inorganic and organic functional groups and works equally well under homonegeous and heterogeneous conditions. The combination of ease of incorporation of azides and alkynes together with the orthogonality of the azide/alkyne *click* reaction has led to a "lego-like" approach to many synthetic routes. The tolerence for other functionality allows the addition of a number of species to the reaction mixture without affecting the CuAAC reaction, including inorganic azides in large excess (Adzima and Bowman, 2012; Binder and Sachsenhofer, 2008; Such et al., 2012).

The advantages of *click* reactions have attracted significant interest from researchers in different fields. Already in 2011, only a decade after its introduction, more than 1500 articles had been published on the further development or implementation of *click* chemistry (Adzima and Bowman, 2012). Besides its starting point in the field of drug design (reviewed by e.g. Lallana et al., 2012), *click* reactions are also useful in polymer chemistry because they are highly efficient (especially azide/alkyne reactions), display a high functional group tolerance, solvent insensitivity, and work equally well under homogenous and heterogeneous conditions (Binder and Sachsenhofer, 2008). *Click* reactions have also proved to be a promising tool in colloidal and surface science (Lutz and Zarafshani, 2008; Nebhani and Barner-Kowollik, 2009). For example, the design of bifunctional molecules that combine a robust anchor that can bind to a variety of metal oxide surfaces and a second functional group that leads to covalent attachment of molecules via *click* chemistry has been proposed as a strategy to introduce a broad range of functionalities onto the surfaces of metal oxide nanoparticles (White et al., 2007). The strategy is applicable to a tremendous diversity of molecules including functional polymers, biomolecules, fluorescent dyes and molecular anchors. Finally, a small but increasing number of studies have been conducted in the field of chemical and biological engineering on account of the large potential benefits from simplification of process design (reaction and separation steps) and resulting cost reduction (Adzima and Bowman, 2012).

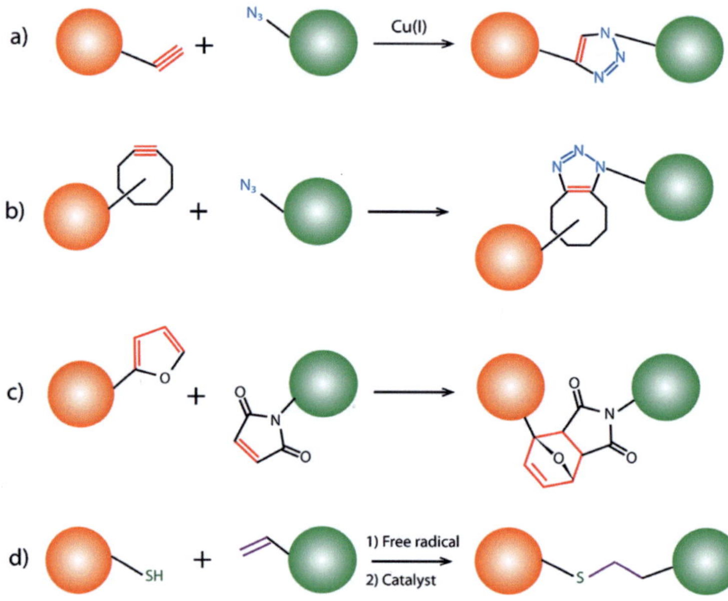


**Figure 2.21:** Selection of reactions that best meet the criteria for a *click* reaction: a) Cu(I)-catalyzed Huisgen 1,3-dipolar cycloaddition of alkynes and azides; b) strain-promoted cycloaddition of alkynes and azides; c) Diels-Alder reaction; and d) thiol-ene reaction, taken from Such et al. (2012) and used with permission from Elsevier.

# Chapter 3

# Materials and methods

A wide range of materials, experimental methods and analytical techniques were necessary to synthesize and characterize the metal oxide and hybrid metal oxide nanoparticles described in this thesis. In this chapter, a full listing of chemicals, experimental and analytical procedures is given. Furthermore, the principles behind the main analytical techniques are explained in this chapter. A detailed description of the pilot plant and its operation is provided in Chapter 4.

## 3.1 Chemicals

| Chemical/Material | Formula | Grade | Source |
|---|---|---|---|
| acetic acid | $CH_3COOH$ | p.a. | Roth |
| acetone | $C_3H_6O$ | technical grade | Merck |
| acetonitrile | $CH_3CN$ | HPLC | Roth |
| ammonium hydroxide solution | $NH_3$ | ACS reagent | Sigma-Aldrich |
| ammonium iron(III) citrate | $C_6H_8O_7 \cdot xFe\ yNH_3$ | purum, iron 20.5 - 23% | Fluka |
| L(+)-ascorbic acid sodium salt | $C_6H_7NaO_6$ | ≥99% | Sigma |
| benzyl azide | $C_7H_7N_3$ | 94% | Alfa Aesar |
| 3-bromo-1-propanol | $Br(CH_2)_3OH$ | 97% | Aldrich |

| Chemical/Material | Formula | Grade | Source |
|---|---|---|---|
| cerium in nitric acid | Ce in $HNO_3$ | standard for ICP | Fluka |
| cerium(III) nitrate hexahydrate | $Ce(NO_3)_3 \cdot 6H_2O$ | 99.99% | Aldrich |
| copper(II) sulfate pentahydrate | $CuSO_4 \cdot 5H_2O$ | p.a. | Sigma-Aldrich |
| *N,N′*-dicyclohexylcarbodiimide | $(C_6H_{11})_2CN_2$ | 99% | Aldrich |
| 4-(dimethyl-amino)-pyridin | $C_7H_{10}N_2$ | 99% | Aldrich |
| dichloromethane | $CH_2Cl_2$ | HPLC | Acros |
| diethyl ether | $(C_2H_5)_2O$ | technical grade | Merck |
| dimethyl sulfoxide-d$_6$ | $(CD_3)_2SO$ | NMR solvent | Aldrich |
| dimethyl sulfoxide | $(CH_3)_2SO$ | p.a. | Merck |
| dodecane | $C_{12}H_{26}$ | gradient cap fluid | CPS Instruments |
| ethanol | $C_2H_5OH$ | ≥99.9% | Roth |
| hexanoic acid | $C_5H_{11}COOH$ | ≥99.5% | Aldrich |
| 5-hexynoic acid | $C_5H_7COOH$ | 97% | ABCR |
| hydrochloric acid | HCl | ≥ 32%, p.a. | Roth |
| iron(III) chloride solution | $FeCl_3$ | ASS | Fluka |
| iron(III) nitrate nonahydrate | $Fe(NO_3)_3 \cdot 9H_2O$ | 99.99% | Aldrich |
| iron(II) acetate | $Fe(CH_3COO)_2$ | 99.99% | Aldrich |
| iron(II) acetate | $Fe(CH_3COO)_2$ | iron, 29.5% min | Alfa Aesar |
| iron oxide(II,III), magnetic nanoparticle solution | $Fe_3O_4$ in $H_2O$ | 5 nm | Aldrich |

| Chemical/Material | Formula | Grade | Source |
|---|---|---|---|
| iron(III) oxide, nanopowder | $Fe_2O_3$ | <50 nm | Aldrich |
| iron(III) oxide, dispersion (red) | $Fe_2O_3$ | 35% in $H_2O$ | NanoAmor |
| iron(III) oxide, powder | $Fe_2O_3$ | <5 $\mu$m | Aldrich |
| iron(II,III) oxide, nanopowder | $Fe_3O_4$ | <50 nm | Aldrich |
| iron(II,III) oxide, powder | $Fe_3O_4$ | <5 µm | Aldrich |
| nitric acid | $HNO_3$ | $\geq$ 65%, p.a. | Roth |
| PVC calibration standards in DI water | | for disc centrifuge | CPS Instruments |
| Rhodamine B | $C_{28}H_{31}ClN_2O_3$ | for fluorescence | Sigma |
| sodium chloride | NaCl | $\geq$ 99.5%, p.a. | Roth |
| sodium hydroxide | NaOH | $\geq$ 99%, p.a. | Roth |
| sodium sulfate | $Na_2SO_4$ | | Merck |
| sucrose | $C_{12}H_{22}O_{11}$ | for microbiology | Merck |
| tetrahydrofuran | $C_4H_8O$ | HPLC | Roth |
| toluene | $C_7H_8$ | 99.9% | Sigma-Aldrich |
| undecanoic acid | $C_{10}H_{21}COOH$ | 98% | Aldrich |
| 10-undecynoic acid | $C_{10}H_{17}COOH$ | 97% | ABCR |

# 3.2 Main analytical methods

## 3.2.1 Differential centrifugal sedimentation (DCS)

Differential centrifugal sedimentation* for the determination of particles size distributions of colloidal systems has a long history. It is based on the phenomenon of sedimentation of particles in a fluid, quantitatively described for spherical particles by applying Stokes' law:

$$D = \sqrt{\frac{18\eta h}{(\rho_\mathrm{p} - \rho_\mathrm{f})\, g t}} \tag{3.1}$$

where $D$ is the diameter of a particle of density $\rho_p$ falling through a fluid of density $\rho_f$ and viscosity $\eta$; the sedimentation distance is $h$ and this distance is covered in time $t$ under the influence of an acceleration $g$. The equation above is applicable for slow-moving spherical particles in dilute dispersions.

When $g$ is the gravitational force, sedimentation is limited to relatively large particles for practical (reasonably short analysis times) and analytical reasons (Brownian motion is negligible). Moreover, particles with $D < 0.1\ \mu$m will not settle by gravity unless they are extremely dense. This limitation can be removed by using (ultra-)centrifuges, which provide radial accelerations up to $10^5\ g$ and allow the sedimentation of particles with $D < 0.01\ \mu$m in reasonable times. Combining a centrifuge with a suitable detector, it becomes possible to measure particle size distributions with high resolution in the nano regime. When using a disc centrifuge (see Figure 3.1 for a schematic drawing), Equation 3.1 changes to:

$$D = \sqrt{\frac{18\eta \ln \frac{R_\mathrm{f}}{R_0}}{(\rho_\mathrm{p} - \rho_\mathrm{f})\, \omega^2 t}} \tag{3.2}$$

where $R_0$ is the position (radius) of the centrifuge at which particles are initially placed, $R_\mathrm{f}$ the position (radius) where the detector is located and $\omega$ is the angular velocity of the centrifuge. Note that if the measured particles are not exactly spherical, $D$ is the Stokes equivalent diameter. This diameter is smaller than the diameter of a sphere of the same mass as the friction (factor) is increased (viscous resistance), e.g. rod like particles with an aspect ratio of $\approx 2$ would appear to be about 5% smaller.

*The background on differential centrifugal sedimentation is based on Bell et al. (2012); Lagaly (2007); Laidlaw and Steinmetz (2005).

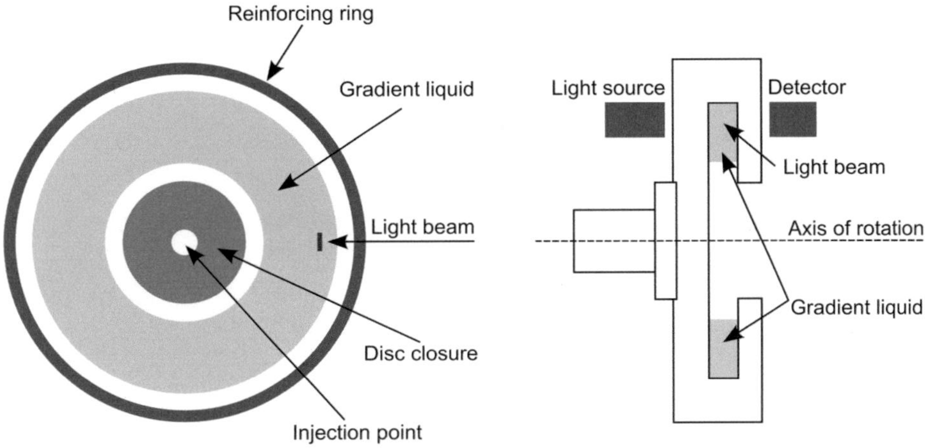


**Figure 3.1:** Scheme of a disc centrifuge for differential centrifugal sedimentation analysis. The drawing is adopted from the scheme shown in Laidlaw and Steinmetz (2005).

For differential sedimentation, the sample is placed on top of a column of clear liquid at the start of the analysis. The detector reports the intensity of a beam of either monochromatic light or X-rays passing through the fluid, initially without particles (maximum intensity). This intensity is reduced as particles reach the detector beam. The reduction is proportional to particle concentration in the case of X-rays, but must be converted to particle concentration by applying the Mie theory in the case of a laser beam. For this calculation, knowledge of the complex refractive index of the particles is of essential importance. The detector signal returns to the original level when all particles have passed the detector. The obtained plot of particle concentration against the calculated particle diameter (using Equation 3.2) produces a differential distribution which may be integrated with respect to particle size to yield an integral distribution.

To run a differential sedimentation analysis, it is important to eliminate the so-called "streaming" or "sedimentation instability", which means that the individual particles do not settle individually according to Stokes' law but rather the sample dispersion settles as a bulk fluid (of slightly greater density) through the liquid column. This can be eliminated by introducing a density gradient to the liquid column. This eliminates streaming by ensuring that the net density of the fluid (being the average density of the fluid plus any suspended particles) increases from top to bottom in the fluid column, thus eliminating the driving force for bulk settling. The condition for eliminating "streaming" may thus be written as

$$\frac{\delta \rho_{net}}{\delta R} \geq 0 \qquad (3.3)$$

where $\rho_{net}$ is the net fluid density and R is the distance from the center of rotation. By ensuring that particles sediment according to Stokes' law, the particles that arrive at the instrument's detector beam have already fractionated so that the detector beam measures only a small slice (a "differential") of the whole size distribution. As the beam width is not infinitesimally small, the sample thickness has a finite size and the distance travelled is also finite, the theoretical resolution is dependant on all these variables. Nevertheless, particle size distributions with very high resolution and high reproducibility can be obtained which is the key strength of this technique. In particular for particles in the size range from 3 nm-10 $\mu$m, this technique allows the resolution of narrow distributions differing by less than 2% and can therefore be used to measure extremely small differences or changes in particle sizes.

In this work, the size distribution of the dispersed particles was determined using a disc centrifuge (CPS Instruments, Florida, USA). The density gradient was established with aqueous sucrose solutions (8%–24% in nine steps) and adjusted to pH 3 using hydrochloric acid for the measurements of unfunctionalized iron oxide dispersions. The disc was spun at constant speed in the range from 18 000 to 24 000 rpm. A laser with a wavelength of 470 nm was always in use. A calibration standard (PVC-particles with a mode of 377 nm) was used prior to each measurement. The first run after the establishment of the gradient was always omitted. To obtain a size distribution based on volume, the knowledge of certain material properties is required. Because of the difficulty of identifying the exact structure of the particles as well as variable properties of iron oxide nanoparticles, obtaining a mass-based particle size distribution is non-trivial. The basis for the optical parameters used in the calculations (real part/imaginary part of refractive index: 2.96/0.5) were those for bulk hematite reported in the literature (Glasscock et al., 2008; Hellwege and Hellwege, 1962; Longtin et al., 1988). In the case of particles synthesized from iron(III) nitrate the bulk density of hematite ($\rho$ = 5.26 g/cm$^3$) was used, whereas the density of ferrihydrite ($\rho$ = 3.96 g/cm$^3$) was used for particles synthesized from iron(II) acetate or mixtures of iron(II) acetate and iron(III) nitrate; both values were taken from Cornell and Schwertmann (2003). Furthermore, a value of 1.1 was used for the non-sphericity factor ($s$) for all calculations. Note that the non-sphericity factor has little effect on the light scattering calculations, for particles that are much smaller than the wavelength used.

Parameter studies were carried out for three different characteristic shapes of particle size distributions observed within this work. The cumulative and differential particle size distributions based on mass ($Q_3$, $q_3$) are shown in Appendix E for all variations considered. Based on the parameter studies, optical and material parameters were chosen conservatively so that the fraction containing larger particles and the size of the particles is more likely to be over than underestimated. Note that trends and the order of magnitude are not influenced by the choice of parameters. In the case of iron nitrate as starting solution, each sample was measured at least three times. For each sample, the mean and the standard deviation was determined from the independent measurements. Due to the very high reproducibility of the measurements and the large number of samples for the investigation on changing the composition, only two measurements in the order of ABCCBA (A, B, C represent a sample) were undertaken. The standard deviation for this procedure was determined from the measurements of characteristic samples.

For the characterization of the dispersed iron oxide nanoparticles synthesized in the presence of 10-undecynoic acid (see Chapter 7), the density of the particles was assumed to be 4.87 g $cm^{-3}$ (density of maghemite). To obtain a size distribution based on volume, a refractive index of 2.5 and an absorption value of 1 was assumed, based on Tepper et al. (2004).

The following characteristic values were calculated based on the particle size distributions (based on mass): mode ($x_{k,3}$), median ($x_{50,3}$), "$y$"-percentile ($x_{y,3}$ with $y$ = 10, 16, 84, 90), geometric mean ($x_{g,3}$), geometric standard deviation ($\sigma_{g,3}$). In addition, the dispersity ($\kappa$) according to the "VDI-Richtlinie 3491" was calculated, defined as

$$\kappa = \frac{x_{84,3} - x_{16,3}}{2x_{50,3}} \tag{3.4}$$

On the basis of this value, particle size distributions are classified as monodisperse for $\kappa < 0.14$, quasi-monodisperse for $0.14 \leq \kappa \leq 0.41$, and polydisperse for $\kappa > 0.41$.

### 3.2.2 Electrophoretic light scattering

Electrophoretic light scattering[†] is a common method for investigations on the colloidal stability of dispersions. The underlying physical principle is that of electrophoresis. A dispersion is exposed to an electric field and the charged particles will migrate toward the oppositely charged electrode (having a non zero electrical potential at the shear plane called the $\zeta$ potential, see below) with a velocity known as the electrophoretic mobility. The velocity is measured by the laser-doppler technique.

Conversion of mobility data to $\zeta$ potentials and especially to the surface charge of the particles is not without arbitrariness, the problem being that the $\zeta$ potential is a potential "close" to the surface of the particles but "close" is not quantitatively defined. However, the $\zeta$ potential is a more intuitive and therefore widely used quantity in collidal chemistry with many applications (in particular, the stability of electrostatically-stabilized dispersions is directly related to the $\zeta$ potential of the dispersed particles). To understand the problems involved in these conversions, it is necessary to examine the interface of a charged particle in a liquid, as shown in Figure 3.2.

Distances within the double layer are considered large or small in relation to the Debye length. In dilute electrolyte solutions, in which the Debye length is large[‡], the surface of shear is very close to the particle surface and may in the limit be assumed to coincide with it. The Hückel equation establishes the relation between electrophoretic mobility and $\zeta$ potential for this case assuming spherical particles. At the other limit, where the magnitute of the Debye length is negligible compared to the radius of curvature of the surface (which practically means high electrolyte concentrations and/or flat or slightly curved surfaces), the relation between electrophoretic mobility and $\zeta$ potential is given by the Helmholtz-Smoluchowski equation. Obviously many systems of interest in colloidal chemistry do not correspond to either of these limiting cases. Therefore, it is recommended to report the electrokinetic mobility instead or at least in addition to a possibly far off $\zeta$ potential value.

In this work, selected hydrothermally synthesized metal oxide dispersions were examined by electrophoretic light scattering using a Zetasizer Nano ZS (Malvern Instruments, Worcestershire, UK) fitted with a 633 nm He-Ne laser (4 mW) and using folded capillary cells with gold electrodes (type DTS1060) to hold the samples.

[†]The background on electrophoretic light scattering is based on Hiemenz and Rajagopalan (1997); Lagaly (2007); Malvern Instruments (2012a,b).

[‡]e.g. for a 0.001 mol $L^{-1}$ aqueous solution of a 1:1 electrolyte, the Debye length is $\approx$ 10 nm

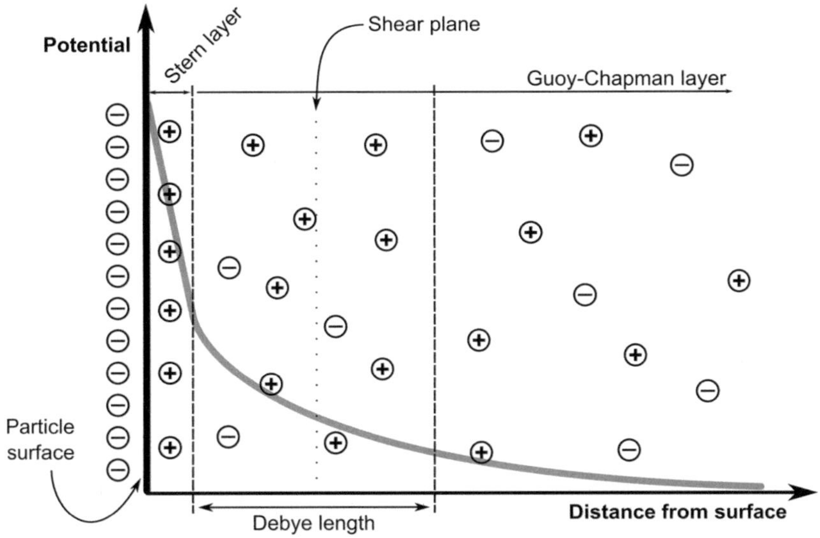


**Figure 3.2:** Schematic depiction of the electrical double layer formed at the interface between a dispersed negatively charged particle and the surrounding medium in the presence of an electrolyte. The Debye length is a function of the ionic strength and the dielectric constant of the dispersion medium. The position of the shear plane is "close" to the surface of the particle but, in general, only qualitatively defined. The potential at the shear plane is defined as the $\zeta$ potential.

## 3.2.3 Raman spectroscopy

Raman spectroscopy§ is a widely and increasingly used method to gain information on the structure, symmetry, electronic environment and bonding of the investigated molecules. It is based on the Raman effect—the inelastic scattering of light which is accompanied by a shift in photon frequency. The shift can be towards higher (anti-Stokes) or lower (Stokes) frequencies depending on whether the photon gains or loses energy from its interaction with the molecule. Although the Raman effect was discovered in 1928, it is only comparatively recently that Raman spectroscopy has emerged as a powerful analytical technique in many fields due to advancements in laser technology, by the design of very efficient filters to suppress the elastically scattered light,

§The general background on Raman spectroscopy is based on Das and Agrawal (2011); Gremlich (2000); Popović et al. (2011); Schmitt and Popp (2006); Schrader (1983)

and by the development of more sensitive detectors. Raman spectroscopy is closely related to infrared spectroscopy and both techniques, when applied to the same sample, provide complementary information. However, the sensitivity of Raman spectroscopy is comparatively poor because only a small number of the incident photons are inelastically scattered. Nevertheless, Raman spectroscopy is better suited than infrared spectroscopy for the examination of aqueous samples, colored substances exhibiting the so-called Raman resonance effect, single crystals and crystal powders, and from microscopic samples such as single fibers or dust grains.

It is possible to take Raman spectra from samples in all aggregation states and the analysis is usually non-destructive. Furthermore, transparent and non-transparent samples can be measured as well as samples with different surface textures and samples contained in closed but optically-accessible vessels (e.g. a quarz cuvette), all without the need for sample pre-treatment. Raman spectroscopy can also be applied to the examination of nanoparticles. Here, Raman scattering enables an indirect determination of nanoparticle properties by analyzing the phonon lines shift and broadening for the particles less than about 30 nm in size. This analysis provides information on the average particle size and size distribution, as well as additional information about phonon confinement, strain and substitutional effects, defect states and non-stoichiometry, electron-phonon coupling etc. in different kinds of nanomaterials.

Concerning iron oxides, several investigations studying the vibrational properties of iron oxides and iron oxyhydroxides by means of Raman spectroscopy were conducted already in the 1970s (Beattie and Gilson, 1970; Reid et al., 1977; Thibeau et al., 1978). The Raman technique was used as a highly sensitive spectroscopic method for the in situ characterization of corrosion products on metal film surfaces in gaseous and aqueous environments (Reid et al., 1977). Nevertheless, the co-existence of different ferric oxide and oxyhydroxide phases, the sometimes non-stoichiometric composition of the iron oxides and oxyhydroxides as well as laser-induced thermal effects and phase changes remain challenges in interpreting recorded Raman spectra correctly (Bersani et al., 1999). More recently, a few studies have reported on detailed investigations with synthetic and natural iron oxide and oxyhydroxides revealing that Raman spectroscopy has the potential to rapidly dinstinguish between the numerous iron oxides and oxyhydroxides when sufficiently low laser powers are applied to avoid phase transformation upon laser irradiation (Chernyshova et al., 2007; Hanesch, 2009; Jubb and Allen, 2010). Furthermore, insights into structural transformations of hematite with decreasing particle size could be gained (destabilization of the hematite phase in the nanoscale regime), but it was found that the structure of hematite nanoparticles was significantly stabilized in a water matrix due to the surface energy loss provided by the electrochemical double layer effect and the increased rigidity of the particle surroundings (Chernyshova et al., 2007). These recent reports—from Hanesch (2009), Jubb and Allen (2010) and especially Chernyshova et al. (2007)—were the starting point for the investigations by means of Raman spectroscopy reported in this work. To identify the structure of the particles directly in the aqueous dispersion, a new measurement setup for performing Raman spectroscopic investigations directly in the aqueous phase was developed.

Raman measurements were performed at room temperature using a Labram HR 800 Horiba Jobin Yvon spectrometer with a charge-coupled device (1024 x 256 pixels) detector. Raman scattering was excited with the 532 nm (2.33 eV) line of a frequency-doubled Nd:YAG laser. Moreover, Raman spectra were also obtained using the 647 nm line of a $Kr^+$ laser. The laser beam was focused on the solid samples investigated by means of a 20x Nikon microscope objective (numerical aperture of 0.35). In order to avoid laser-induced heating or thermal decomposition of the specimen the applied laser power density was carefully adjusted by a filter and kept very low (850 W $cm^{-2}$). That is why acceptable Raman spectra were obtained within 5-10 minutes. Applying the following sampling technique (see Figure 3.3) aqueous suspensions of iron oxide nanoparticles were characterized in situ by means of Raman spectroscopy: In contrary to Chernyshova et al. (2007) the aqueous suspension of iron oxide nanoparticles was prepared as 'hanging drop' on a cover glass which was positioned above a cavity of the sample holder. As a consequence, the evaporation of the sample droplet is minimized due to the enclosed chamber allowing a detailed examination of ferrous oxide nanoparticles in suspension. By passing the laser beam through a 100x Olympus microscope objective (numerical aperture of 0.80) with long working distance the focal point was set 10 μm below the interface between cover glass and sample droplet. The applied laser power density was in the range of 50 to 100 kW $cm^{-2}$. Investigating solid specimen and aqueous suspensions, in both cases the scattered light collected in backscattering geometry by the used 20x or 100x microscope objective, respectively, was not polarization analyzed allowing both parallel and perpendicular components to be detected simultaneously.

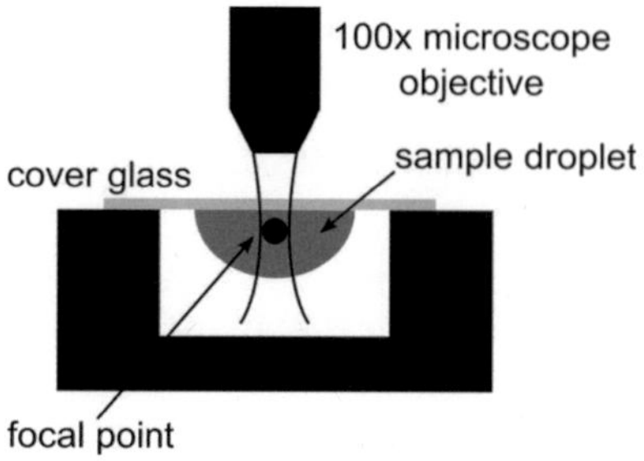


**Figure 3.3:** Schematic drawing (cross section) of the Raman setup investigating aqueous suspensions of iron oxide nanoparticles: The suspension was prepared as "hanging drop" on a cover glass positioned above a cavity of the sample holder. The laser was focused by means of an 100x microscope objective 10 μm below the interface between cover glass and sample droplet. Drawing kindly provided by C. Röder, Institut für Theoretische Physik, TU Bergakademie Freiberg.

### 3.2.4 Powder X-ray diffraction (PXRD)

X-ray diffraction (XRD)¶ is an invaluable and mostly non-destructive technique for the analysis of crystalline structures. In general, diffraction occurs when waves encounter a series of regularly spaced obstacles that are capable of scattering the waves and have spacings that are comparable in magnitude to the wavelength. Diffraction patterns arise as a consequence of interference (constructive and destructive) between waves that have been scattered by the obstacles. XRD is based on the diffraction of electromagnetic radiation with wavelength ranging from a few Ångströms to 0.1 Ångström (hard X-rays), which is in the order of the atomic spacings of solids. The conditions for constructive interference and consequently, the emergence of a diffraction pattern are illustrated in Figure 3.4: If two waves scatter off two parallel planes of atoms separated by a distance $d$, constructive interference will occur when the distances travelled by the waves between source, scattering planes and detector differ by an integer multiple ($n$) of the wavelength ($\lambda$). This relationship is given by Bragg's law:

$$n\lambda = 2d \sin\theta \tag{3.5}$$

where $\theta$ is the angle between the incident waves and the planes of atoms.

¶The background on XRD and PXRD is based on Callister (1997); Laidler and Meiser (1995); Paulus and Gieren (2001); Serway (1996).

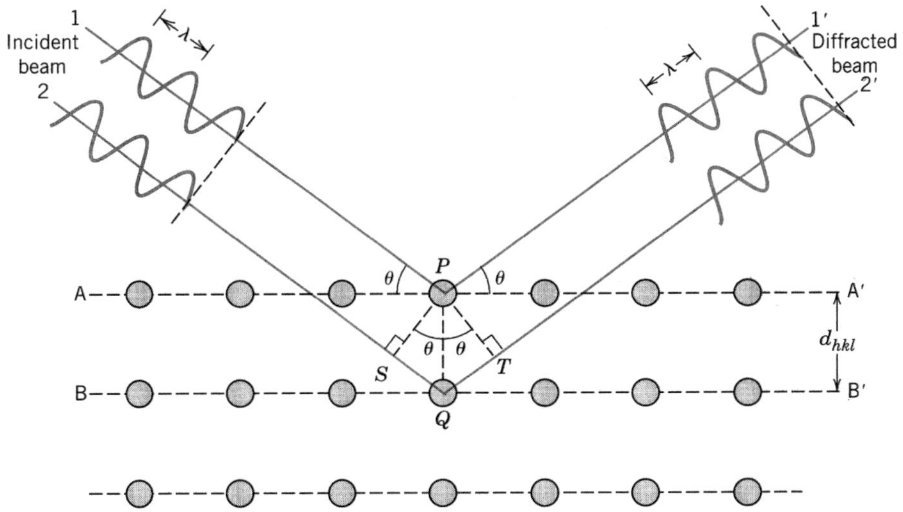


**Figure 3.4:** Diffraction of X-rays by parallel planes of atoms. Depicted are two waves (1 and 2) of equal wavelength $\lambda$ emerging from an X-ray source ("incident beam") and diffracting off parallel planes (A and B) separated by a distance $d$ ($d_{hkl}$ in the figure). If P and Q are the scattering centers at the two planes, wave 2 (striking the lower plane) travels a distance $\overline{SQ} + \overline{QT}$ longer than wave 1. Constructive interference occurs when this distance is an integer multiple of the wavelength. Figure is taken from Callister (1997).

This is the basis for the determination of the interplanar spacings of the investigated crystals and the reconstruction of the unit cell size and shape. Furthermore, the energetic X-rays can penetrate quite deep in to the materials and therefore provide information about the bulk structure. The typical absorption lenghts range from some millimeters to few microns, depending on the wavelength of the X-ray and on the material used in the analysis. X-ray scattering is most sensitive to high atomic number elements (due to the higher electronic density).

Several X-ray techniques have been developed for structural investigations of matter. One widespread diffraction technique employs powdered or polycrystalline specimens consisting of many fine and randomly oriented particles that are exposed to monochromatic X-rays (Powder X-ray Diffraction, PXRD). Here a large number of randomly oriented crystals are present, ensuring that some particles are properly oriented such that every possible set of crystallographic planes will be available for diffraction. The apparatus used determines the angles at which diffraction occurs for the "powdered" sample.

Besides the exact peak positions in the diffraction diagram, the width of the peaks contains valuable information on the investigated material. The finite crystallite sizes as well as structural defects affect the diffraction pattern and contribute to the broadening of the diffraction peaks. Different data processing methods exist to descriminate between the different influences, starting from the fairly simple Scherrer equation to determine the crystallite size by using the half width of a single peak to much more advanced methods using a whole pattern fit.

In the work reported here, PXRD measurements of the dried particles were performed with a Bragg-Brentano-Goniometer D8 (Bruker AXS) using monochromatic Cu K$\alpha$ radiation, a secondary monochromator, sample changer, and a scintillation counter. The volume weighted crystallite sizes were determined from the Lorentz widths, of the diffraction peaks. Correction for the instrument broadening was performed using the standard reference material 660a ($LaB_6$) of the National Institute of Standards & Technology (NIST). The crystalline phases present in the samples were identified by comparing measured positions and relative intensities with the reference lines of relevant iron oxide phases from the International Centre for Diffraction Data (ICDD, USA).The composition of the iron oxide phases was calculated by applying quantitative Rietveld analysis on the basis of structure data reported in literature.

### 3.2.5 Transmission electron microscopy (TEM)

Electron microscopy[‖] is a powerful method to provide magnified images of objects at a much higher spatial resolution than optical microscopy, allowing detailed micro- and nanostructural examinations. Furthermore, this electron beam based imaging techiques are usually coupled to other characterization methods providing elemental and chemical analysis by e.g. energy-dispersive X-ray spectroscopy (EDX). The two types of electron microscopy techniques used in this work were transmission electron

[‖] The background on electron microscopy is based on Callister (1997); Gatelli et al. (2004); Kriete et al. (2005); Serway (1996); Shackelford (1985)

microscopy (TEM) and the closely related scanning transmission electron microscopy (STEM).

The arrangement of a transmission electron microscope is similar to that of an optical microscope but accelerated electrons are used instead of visual light and solenoid coils play the part of "magnetic lenses" instead of optical lenses. Because the accelerated electrons interact strongly with atoms by elastic and inelastic scattering, the sample must be very thin. In TEM, the image is built through the differences in scattering or diffraction produced by the various features of the sample. In STEM, the electron beam is focused and scanned over the sample in a raster, allowing for additional analyses such as EDX to be performed on the sample (analogous to scanning electron microscopy, SEM). EDX analysis is possible because the interaction of the accelerated electrons with the sample produces X-rays with wavelengths characteristic of the atoms present in the sample. Therefore, STEM allows for simultaneous imaging and elemental characterization of the samples.

Transmission electron microscopy (TEM) analysis was performed using either a FEI CM 200 FEG microscope or an image corrected FEI Titan 80-300 microscope (Hillsborough, USA) operated at 300 kV and equipped with a Gatan US1000 CCD camera for TEM imaging. Scanning transmission electron microscopy (STEM) images were acquired using a HAADF (High Angle Annular Dark Field) detector with a nominal spot size of 0.14 nm. Energy dispersive X-ray (EDX) spectra were obtained with a EDAX s-UTW detector in STEM mode with a beam width of 0.3-1 nm. When the first microscope was used, the particle suspension was nebulized via ultrasound on Lacey grids. When the second microscope was used, TEM samples were prepared by drop-casting diluted suspensions of the nanocrystals onto carbon coated Cu grids (Quantifoil holey carbon grids coated with 2 nm carbon). Excess solvent was taken off after 30 sec and the grids were rinsed with ethanol and dried at room temperature.

### 3.2.6 X-ray photoelectron spectroscopy (XPS)

X-ray photoelectron spectroscopy** is a widely used method to analyze the surface chemistry of materials. It is based on the photoelectric effect: a photon of sufficiently high energy $h\nu$ can eject an electron (photoelectron) from an atom. In XPS, soft X-rays

**The background on XPS is based on Rivière et al. (2011) and Gatelli et al. (2004) which itself draws from Ashcroft and Mermin (1976); Brundle et al. (1992); Cardona and Ley (1978-1979); Eisberg and Resnick (1985); Luth (2001).

of a fixed and narrow wavelength distribution (e.g. Al K$\alpha$ = 1486.6 eV) are used which cause electrons from the core levels of atoms to be ejected. The kinetic energy ($E_{kin}$) of the ejected photoelectron depends on the energy of the photon following

$$E_{kin} = h\nu - E_{bin} - \Phi_S \tag{3.6}$$

where $E_{bin}$ is the binding energy of the electron in the sample and $\Phi_S$ is the spectrometer work function. Therefore, it is possible to calculate the binding energy of electrons in the core levels of atoms by measuring their kinetic energy upon ejection.

Although the binding energy of a core level electron is approximately constant for an atom, changes in either oxidation state of the atom or its chemical and physical environment give rise to small shifts in the peak positions in the XPS spectrum. For example, photoelectrons coming from atoms of a higher positive oxidation state exhibit a higher binding energy due to the extra Coulombic interaction with the ion core. The ability to discriminate between different oxidation states and chemical environments is one of the major stengths of the XPS technique.

Because electrons can only travel short distances through solids without losing energy in inelastic collisions with atoms, the electrons ejected from the sample come from the first atomic layers (the depth is dependant on the material, the energy of the photoelectron and the angle of measurement), giving XPS its surface sensitivity. XPS is able to identify all elements, except hydrogen and helium, with a probing depth of a few tens of monolayers. All XPS measurements are carried out under ultra high-vacuum (UHV) to minimize collisions of the emitted photoelectrons in the path length between sample and detector. Moreover, UHV is necessary due to the fact that at $10^{-6}$ mbar, a monolayer of residual gas can be adsorbed in less than 1 s; considering the XPS sampling depth of 5-10 nm, such contamination layers hamper the desired surface information. A schematic of the XPS-apparatus used in this work (taken from Nunney and White (2011)) is shown in Figure 3.5.

For the work presented here, XPS measurements were performed using a K-Alpha XPS spectrometer (ThermoFisher Scientific, East Grinstead, UK). Data acquisition and processing using the Thermo Avantage software is described in Parry et al. (2006). All nanoparticles were analyzed using a microfocused, monochromated Al K$\alpha$ X-ray source (30-400 $\mu$m spot size). The K-Alpha charge compensation system was employed during analysis, using electrons of 8 eV energy and low-energy argon ions to prevent any localized charge build-up. The spectra were fitted with one or more Voigt profiles (binding energy uncertainty: $\pm$ 0.2 eV). The analyzer transmission function,

Scofield sensitivity factors, and effective attenuation lengths (EALs) for photoelectrons were applied for quantification (Scofield, 1976). EALs were calculated using the standard TPP-2M formalism (Tanuma et al., 1994). All spectra were referenced to the C 1s peak of hydrocarbon at 285.0 eV binding energy controlled by means of the well known photoelectron peaks of metallic Cu, Ag, and Au, respectively.

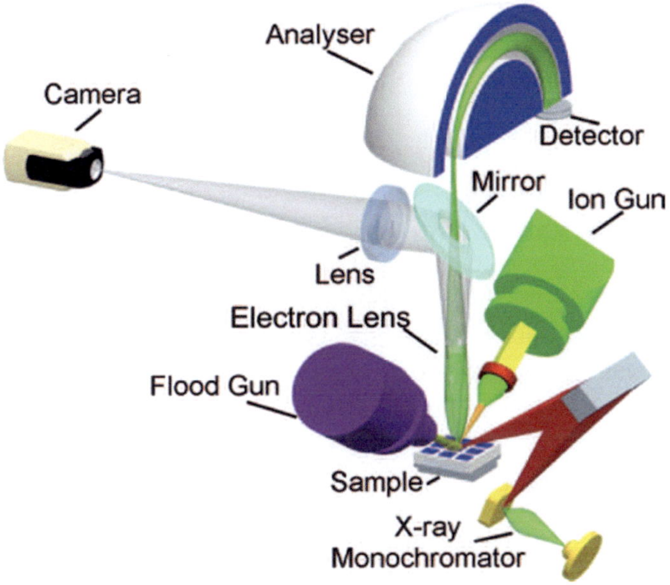


**Figure 3.5:** Sketch of the internal configuration of the XPS-apparatus used within this work (Thermo Scientific K-Alpha). The photoelectrons generated upon monochromatic X-ray irradiation of the sample are collected (electron lens), selected based on kinetic energy (analyser) and counted (detector). A low-energy electron gun (flood gun) for analyzing insulators and an ion gun for cleaning and depth profiling. Figure is taken from Nunney and White (2011).

## 3.3 Experimental procedures

### 3.3.1 Continuous hydrothermal particle synthesis

The equipment and general procedure used to hydrothermally synthesize metal oxide particles in a continuous process are described in detail in Chapter 4.

### 3.3.2 Determination of the conversion of the metal salts

In order to determine the conversion of the metal salts into metal oxide particles, the particles were removed from the hydrothermally synthesized dispersions by means of membrane filtration from a stirred pressure vessel (pressure vessel from Berghof, Eningen, Germany equipped with a PESH-UH050 membrane from Mycrodyn-Nadir, Wiesbaden, Germany). The permeate was analyzed for cerium or iron, respectively, using inductively coupled plasma-optical emission spectroscopy (ICP-OES; Model: iCAP 6300, Thermo Electron GmbH, Dreieich, Germany) and for nitrate using ion chromatography (column: IonPac AS11-HC, Dionex; eluent: 0.3 mol $L^{-1}$ NaOH, detection at 220 nm).

### 3.3.3 Synthesis of iron oxide reference materials

Hematite nanoparticles were prepared by forced hydrolysis of iron(III) salts according to method 1 in Schwertmann and Cornell (2000). However, the dispersion was freeze dried prior to analysis. 6-line ferrihydrite and 2-line ferrihydrite were prepared as described in Schwertmann and Cornell (2000). Iron(II,III) oxide nanopowder (<50 nm particle size (TEM), $\geq$98% trace metal basis) and iron(II, III) oxide powder (<5 μm, 95%) were purchased from Aldrich and used as received. X-ray diffraction patterns and Raman spectra were taken for all samples (results are shown in Figure E.2 on page 222 and in Figure 6.3 on page 103, respectively).

### 3.3.4 Testing the stability of the 5-hexynoic acid and the 10-undecynoic acid under process conditions

The same high pressure pilot plant was used as described for the synthesis of the primary functionalized particles but operated at slightly modified modes (cf. Chapter 4). For testing the stability of the 5-hexynoic acid, a water stream of 100 mL $min^{-1}$ was mixed with a stream of 0.28 mL $min^{-1}$ of 5-hexynoic acid with temperatures of 573 K, 623 K, 673 K, 698 K in the reaction zone. For the stabilty testing of 10-undecynoic acid, the main water stream had a flow rate of 100 mL $min^{-1}$, the ethanolic stream containing the undecynoic acid was delivered with a flow rate of 2 mL $min^{-1}$ and the aqueous stream containing ammonia at 25 mL $min^{-1}$ with a concentration of 0.43 wt% with temperatures of 573 K, 623 K, 673 K in the reaction zone. The product solutions were analyzed for 5-hexynoic acid and for 10-undecynoic acid respectively with ion trap mass spectrometry using electrospray ionization (6300 Agilent LC/MS, Waldbronn, Germany).

### 3.3.5 Preparation of the functionalized particles for characterization

The product dispersions containing particles synthesized in the presence of 5-hexynoic acid and 10-undecynoic acid went through several cycles of centrifugation, decantation and washing with water and then ethanol. After the last decantation, the particles were dried at 333 K for 24 h. Furthermore, for some particle dispersions containing particles synthesized in the presence of 10-undecynoic acid, a different fraction went through cycles of centrifugation, decantation and washing, first with water, then twice with ethanol, and then with a 1 wt% ammonia solution. After this last step, the particle suspensions were placed for 5 min in a sonication bath leading to well dispersed particles. Only a fraction of the particles was sedimented upon centrifugation of the resulting particle suspension. The remaining dark dispersion (supernatant) was stable for several months.

### 3.3.6 Characterization of the functionalized particles

#### Phase transfer test

A fraction of the as received particle dispersions was mixed with toluene (around 1:1 by volume) and the mixture was allowed to phase seperate. The position and the distribution of the coloured particles was recorded as seen by the naked eye.

#### *Click* reaction of primary functionalized particles with the model compound benzylazide

*Click* reactions with the primary functionalized particle were carried out with 20 mg of the particles synthesized in the presence of 5-hexynoic acid which were redispersed using a sonication bath in 20 ml DMSO/$H_2O$ (4:1 by volume) in a round bottom flask. Then, 6.5 mg benzylazide, 12.4 mg $CuSO_4 \cdot 5H_2O$ and 80 mg L(+)-ascorbic acid sodium salt were added and mixed with a shaker for at least 16 h at room temperature. Subsequently the particles went through several cycles of washing, centrifugation and decantation first with DMSO, then with water and thereafter with ethanol, and were dried at 60 °C in a drying chamber. For particles synthesized in the presence of 10-undecynoic acid, a very similar procedure was followed. However, the particles were directly added in suspended form to the mixture containing benzylazide in excess.

#### Synthesis of reference substances

1. Na-salts of carboxylic acids were prepared by dissolving the acids (hexanoic acid, 5-hexynoic acid, undecanoic acid, 10-undecynoic acid) in water with an at least equimolar amount of NaOH. Subsequently the water was removed under reduced pressure.
2. *Click* reaction of alkynoic acids with benzylazide: 464 mg of 5-hexynoic acid (4.14 mmol) was dissolved in 10 mL of DMSO/$H_2O$ (4 : 1 by volume). 568 mg benzylazide (4.14 mmol), 124 mg copper(II) sulfate pentahydrate (0.5 mmol), 162 mg L(+)-ascorbic acid sodium salt (0.82 mmol) was added and the reaction mixture was stirred for 16 h. Subsequently, the 4-(1-benzyl-1H[1,2,3]triazol-4-yl)butyric acid was obtained and purified by a few cycles of addition of water to

precipitate the substance, centrifugation, decantation and dissolution in DMSO. The last precipitate was dried under reduced pressure. $^1$H NMR (400 MHz, DMSO-$d_6$): $\delta_{ppm}$ 1.72 – 1.89 (m, 2H, $CH_2$-$CH_2$-$CH_2$), 2.25 (t, J = 7.07 Hz, 2H, $CH_2$-COOH), 2.62 (t, J = 7.45 Hz, 2H, C-$CH_2$-$CH_2$), 5.54 (s, 2H, Ar-$CH_2$-N), 7.23 – 7.45 (m, 5H, Ar *H*), 7.92 (s, 1H, N-*CH*-C ), 12.07 (br.s., 1H, COO*H*).

### Characterization of the functionalized particles using a fluorescence dye containing an azide group

**Synthesis of $N_3$-Rhodamine** $N_3$-Rhodamine was synthesized as described in Goldmann et al. (2010). Briefly, 3-bromopropan-1-ol (5 g, 36 mmol) and sodium azide (3.83 g, 59 mmol) were dissolved in a mixture of acetone (60 mL) and water (10 mL) in a round-bottomed flask and the solution was refluxed overnight. Acetone was then removed under reduced pressure, 50 mL of water was added, and the mixture was extracted with diethyl ether (3 x 50 mL). The collected organic layers were dried over $Na_2SO_4$ and 3-azidopropan-1-ol was isolated as a colorless oil following the removal of the solvent under reduced pressure. We then dissolved 3-azidopropan-1-ol (2 g, 0.02 mol), Rhodamine B (10.6 g, 0.022 mol), *N,N'*-dicyclohexylcarbodiimide (DCC; 8.24 g, 0.04 mol) and 4-dimethylaminopyridine (DMAP; 0.102 g, 0.84 mmol) in 200 mL dichloromethane. The reaction mixture was stirred overnight (18 h) and following the removal of the solvent, the resulting dark red liquid was purified by flash chromatography (CombiFlash Companion, Teledyne Isco, Inc., NE, U.S.A., RediSep column, acetone: methanol concentration gradient 0-100%). IR spectra before and after the linkage of the azide group are shown in Figure F.1 on page 257.

***Click* reaction between $N_3$-Rhodamine and the primary functionalized particles** About 20 mg of the particles were redispersed in 3 mL DMSO/$H_2O$ (4:1 by volume) using a sonication bath. Subsequently, approx. 6.2 mg copper(II)sulfate pentahydrate, 40 mg L(+)-ascorbic acid sodium salt and 20 mg $N_3$ Rhodamine were added and mixed with a shaker for 48 h at room temperature. Subsequently, the particles went through several cycles of washing, centrifugation and decantation first with DMSO, then with water for the following cycles until the washing solution showed no sign of fluorescence. The particles were redispersed in about 1 mL of water and the last washing solution was kept for investigations with LSCM.

**Preparation and imaging of the particles with LSCM** A droplet of a particle suspension or washing solution was mixed with and therefore subsequently embedded into mounting media (Mowiol, Höchst) between a glass microscope and a cover slide (Roth, Karlsruhe, Germany). Probes were analyzed using a laser scanning confocal microscope (LSM 510 Meta, objective: Plan-Apochromat 63x/1.4 Oil DIC, Zeiss, Jena, Germany). Fluorescence images were obtained at 6% laser power with an excitation wavelength of 561 nm and the fluorescence emission was passed through 572-615 nm filter.

#### Characterization of magnetic properties

The magnetic properties of selected functionalized iron oxide particles were investigated by measuring their magnetization curve using an "Alternating Gradient Magnetometer" (AGM, Princeton Measurements Corporation, Princeton, NJ, USA). Saturation magnetization ($M_s$) and the remanence/remanent magnetization ($M_r$) were derived from the magnetization curves and used as characteristic values for each sample. For these measurements, 2-4 mg samples of dried particles were weighed into glas capillaries, compacted such that no large voids remain, sealed, and introduced into the AGM where they were exposed to an alternating magnetic field from $+1$ T to $-1$ T. Calibration of the system was carried out using a nickel standard of known weight and magnetic properties.

### 3.3.7 Synthesis and characterization of "glyconanoparticles" using galactose with an azide-terminated linker

The galactose with an azide-terminated linker was provided by Dr. Stefan Jennewein from the Fraunhofer Institute for Molecular Biology and Applied Ecology IME. The structure is shown in Figure 3.6.

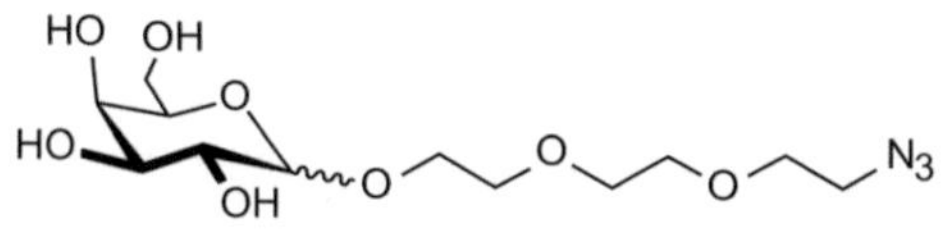


**Figure 3.6:** Structure of galactose with azide-terminated linker.

### *Click* reaction between galactose with an azide-terminated linker and the primary functionalized particles

The *click* reaction procedure was analogous to that described above (Section 3.3.6). The galactose with the azide-terminated linker was provided in excess. After the *click* reaction the particles were separated using a strong magnet and the remaining solution was replaced several times first by DMSO/$H_2O$ (4:1 by volume) and then with decreasing DMSO content, then with $H_2O$ and finally with ethanol. After removal of the last remaining ethanol solution, the particles were dried at 60 °C in an oven.

### Estimation of the increase in organic load on the particles through secondary functionalization

For a rough estimation of the increase in organic load through secondary functionalization, selected samples (with and without secondary functionalization) were subjected to thermogravimetric analysis (TGA) using a TGA Q5000 from TA Instruments (New Castle, Delaware, USA). The samples without secondary functionalization underwent the same reaction and separation procedures as the particles with secondary functionalization but without the addition of the galactose with the azide-terminated linker.

For the TGA measurements, 0.2–2 mg samples of dried particles were weighed into a TGA sample container and heated under a nitrogen gas flow at a rate of 5 K/min from room temperature up to 973 K. The organic loads were estimated from the recorded thermogravimetric curves. In addition, TGA measurements of the stable aqueous dispersion of the in situ functionalized particles (with undecynoic acid, primary functionalization) were carried out. For this, approx. 15 mg of the dispersion was directly filled into the sample container and weighed. TGA was performed also at a rate of 5 K/min from room temperature up to 973 K.

# Chapter 4

# Conceptual design, construction and operation of the pilot plant

The continuous hydrothermal particle formation process, based on the simple principle of mixing a cold metal salt stream with a hot compressed water stream, was carried out in a newly designed and constructed high pressure pilot plant. In this chapter, the conceptual design, the construction, and the operation of the pilot plant are described in detail.

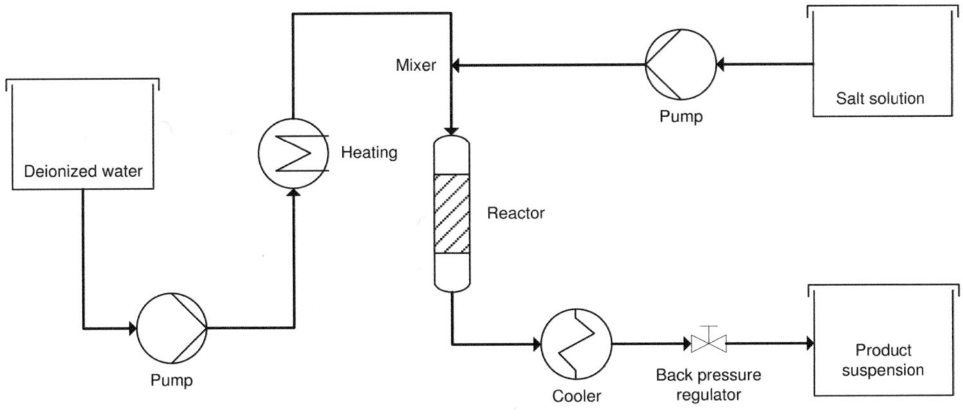


**Figure 4.1:** Simplified process flow diagram of the pilot plant for the hydrothermal synthesis of metal oxide nanoparticles.

## 4.1 Detailed design of the high pressure pilot plant

A simplified flow diagram containing the essential parts of the high pressure plant is shown in Figure 4.1. A photograph of the high pressure pilot plant is shown in Figure 4.2 and a detailed flow diagram in Figure C.4 on page 212.

**Figure 4.2:** Photograph of the high pressure pilot plant.

Generally, deionized water is transported with a high pressure pump and heated appropriately in order to obtain a certain mixing temperature upon mixing with the cold metal salt stream. After passing the reactor the stream is cooled and recovered as a particle suspension. The pressure is controlled by a back pressure regulator.

In the detailed design of the high pressure pilot plant, special emphasis was given to the following aspects:

1. realization of high total mass flow rates in comparison to the pilot plants used by other research groups
2. easy adjustment of the mixing and reactor geometry
3. low pulsation

The first point enables very short residence times and the examination of flow velocity on particle formation over a very wide range of mean flow velocities. The second point is required to study the influence of different mixing geometries, to be able to introduce additional streams at different positions in the reaction zone, and to be able to vary the residence time at the same total mass flow rates. The third point was necessary because of the sensitivity of the reaction conditions to even small changes in pressure. Details on certain aspects of the pilot plant are given below.

### 4.1.1 Pumps

Several high pressure pumps for the transport of different streams were needed for the realization of different configurations (see Section 4.2). Concerning the main water stream, in most cases two coupled syringe pumps (Model 500 HP, high pressure version, Teledyne Isco., USA) were in use which are by their design principle of practically no pulsation. Up to about 150 ml $min^{-1}$ could be continuously delivered at a maximum pressure of 34.4 MPa using these pumps. Alternatively, an HPLC-pump specially designed to deliver fluids also with practically no pulsation in high precision within a very broad range of flow rates was used (SD1, AlphaCrom OHG, Langenau, Germany; for the version used within this work: from 0.01 to 200 ml/min at a pressure of 27.5 MPa and up to around 180 ml/min at 30 MPa).

Up to three HPLC-pumps (PU 2080, $p_{max}$ = 50 MPa, Jasco, Gross-Umstadt, Germany) were initially used to deliver the metal salt solutions with a flow rate of up to 10 ml/min (each) with low pulsation. However, using these pumps at maximum speed with aqueous metal salt solutions rapidly caused abrasion of the piston seal, hindering reliable long term experiments. Therefore, also for the transport of the aqueous metal salt solutions either two coupled syringe pumps (Model 260 D, Teledyne Isco) or a second specially designed HPLC-pump (SD1, AlphaCrom) proved to be much more reliable and practical. However, for the transportation of solutions consisting of or containing organic molecules, the small HPLC-pumps (PU 2080) could be reliably used. The exception was the ethanolic stream containing the 10-undecynoic acid which was delivered using a SD1 pump. A double diaphragm metering pump (EK2, $p_{max}$ = 45 MPa, Lewa, Leonberg, Germany) with a pulsation damper or one of the SD1 HPLC-pumps was used for experiments that required the delivery of an aqueous stream into the main water stream just before the cooler. Also a diaphragm metering pump (EH1, $p_{max}$ = 40 MPa, Lewa) was used for backwashing the filter unit and a different double diaphragm metering pump (Cerex Typ N - KH 42, Bran & Luebbe, Norderstedt, Germany) was sometimes used to accelarate the cooling of the pilot plant after carrying out the experimental runs.

### 4.1.2 Heating and cooling

The main water stream was heated to appropriate temperatures by sequentially passing through high pressure tubings wound around electrical heating blocks. The first three custom made (at Fraunhofer ICT) heating blocks were taken from dismantled high pressure pilot plants: the first two consisted each of 24 m of high pressure tubing wound around a heating block made of aluminum with eight inserted heat cartridges (O.D. = 10 mm, length = 130 mm, 315 W, Türk & Hillinger, Tuttlingen, Germany) and the third one consisted of three 6 m high pressure tubings (O.D. = 1/4 in.) each wound around a heating block made out of copper. The last heating unit before the mixing with the cold streams consisted of a 6 m high pressure capillary (O.D. = 1/8 in.) wound around a custom made copper heating block (see Figure C.1 on page 209). The T-union for the insertion of the thermocouple into the reaction zone was embedded in a custom made heating block (see Figure C.2 on page 210) and the main high pressure capillary was passed through a custom made copper zylinder with a through hole made by wire-electro discharge machining (Isis Technolgies, Karlsruhe, Germany, see Figure C.3 on page 211) to ensure that the mixing temperature was maintained throughout the reaction zone. This element was heated with two band heaters (Typ H, I.D. = 1½ in., width = 3 in., 500 W, Watlow, Kronau, Germany). The custom made copper heating units were all chemically nickel-plated with a coating thickness of 10 $\mu$m (RST, Keltern, Germany). The hot part of the pilot plant was highly insulated to minimize heat losses. The three main heating blocks were covered by specially manufactured insulation covers. All other parts were covered with insulation material (Carbowool up to 1300 K with a thickness of 20 mm or 25 mm, Carborundum Insolation Technology) and/or glass wool insulation rolls held by aluminum foil.

After passing through the reaction zone, the combined stream was cooled to a temperature below 298 K either with an external water jacket (the stream passing through 6 m of 1/4 in., $\delta_w$ = 1.65 mm, high pressure tubing) or by direct quenching with a cold water stream for more rapid cooling. At the beginning, another external water jacket identical in construction and placed in series with the other cooler was used to cool the aqueous metal salt stream below 298 K just before entering the hot compressed water stream. However, the distance between the cooler and the entrance into the hot water stream was too far to really influence the temperature of the entering fluid. Therefore, the cooler was later on disconnected.

### 4.1.3 Mixing and reaction geometry

The zone between the first cold stream being delivered into the hot compressed water stream and the entrance of the combined stream into the cooler was designed in such a way that its length could be easily varied by moving the cooler up and down along two guidance rails. Therefore, different mixing units and reaction geometries could be introduced as needed.

Different mixing units were used for the mixing of cold streams with the hot compressed water stream. In the case of the mixing with the aqueous metal salt stream, standard 1/8 in. T-unions were initially used (SS-200-3, Swagelok, B.E.S.T. Fluidsysteme, Karlsruhe, Germany). However, due to an unfavorable mixing geometry (step changes in the I.D. between 1.75 mm and 2.29 mm, see Figure 4.3) with the result of easy clogging of the part of the mixing unit where the hot compressed water stream enters the mixing unit, a different T-union with an invariable I.D. was introduced (720.1633, Sitec-Sieber, Maur, Switzerland, I.D. 1.6 mm). The geometries of these two mixing units are shown in Figure 4.3. To connect the T-union to 1/8 in. capillaries, a small piece of 1/4 in. tubing with the same I.D. as the through holes in the T-union (1.6 mm) was connected via the standard Sitec high pressure connection to the T-union and connected to the 1/8 in. capillary via hard-soldering. When introducing a second stream directly before or after the cold metal salt stream two of these just described T-unions were connected directly with the smallest piece of 1/4 in. tubing possible,with the inlets for the cold streams point in opposite directions. An assembly of the two T-unions with the adjacent tubings and connectors is shown in Figure 4.4.

A standard 1/4 in. T-union (SS-400-3, Swagelok) was used to introduce a cold stream just before the entrance into the external water jacket. The length of the high pressure capillary can be varied to accomplish a very wide range of residence times in the reaction zone at the same mean flow velocity. Depending on the required residence time, the capillary could either be passed through a hole in a heated copper cylinder or could be wound around a different copper zylinder (see Section 4.1.2) for maintaining the reaction temperature. For all experiments described in this thesis, a 1/8 in. capillary of 250 mm in length passed through a 230 mm copper cylinder (refer to Section 4.1.3) was in use.

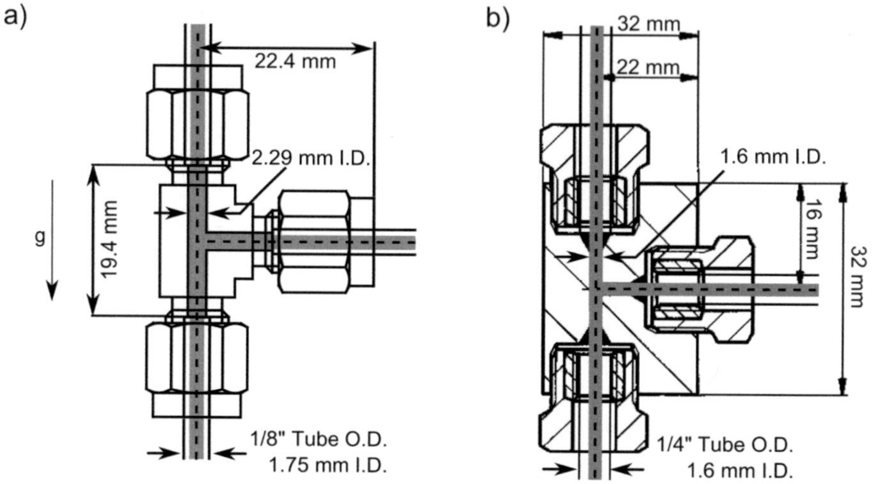


**Figure 4.3:** Geometries of the two used mixing units: a) 1/8 in. T-union from Swagelok, b) 1/4 in. T-union from Sitec.

## 4.1.4 Tubing and fittings

Standard high pressure tubings and fittings out of stainless steel grade 316 (1.4401, components from Swagelok), 316L (1.4435) or 316Ti (1.4571, the latter ones both for components from Sitec) were used for the construction of the pilot plant. All streams joining the main hot compressed water stream were secured against return flow by lift check valves (SS-53S4, Swagelok).

The cooled combined stream was always passed through one of two filter units (TF series, poresize around 10 $\mu$m, Swagelok) to protect the back pressure regulator from deposits of larger particles. In case of the clogging of one of the filters, the stream could be switched by a three-way valve (Typ 5, Sitec-Sieber) to pass through a parallel pathway containing the other filter unit.

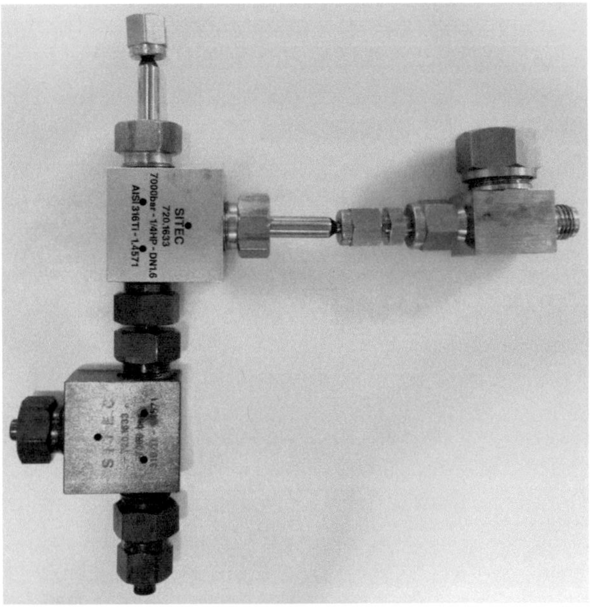


**Figure 4.4:** Picture of a typical assembly of two mixing units with the adjacent tubings and connectors.

### 4.1.5 Process measurement, control and data logging

Special attention had to be given to the precise measurement and control of pressure and temperature due to the sensitivity of both equilibrium and kinetics of the relevant reactions to small changes in these parameters.

Custom made thermocouples were used for carrying out systematic studies on the influence of the mixing and the total mass flow rate on the properties of the product dispersions with the model systems cerium oxide (Chapter 5) and iron oxide (Chapter 6). The thermocouples were designed so that the dead volume through the integration of the T-union to introduce the thermocouple into the stream was minimized and the cross section was barely influenced by using a thermocouple with a small diameter which also reduces the response time (thermocouple type K, O.D. 0.5 mm, Testo, Lenzkirch, precision of the measurement chain: $\pm$ 0.5 K) where in use for critical measuring points. This applied to the measurement of the temperature of the hot compressed water stream ($T_{\text{hot water}}$) and the cold metal salt stream ($T_{\text{salt}}$) just before their entrance into the mixing unit and to the measurement of the temperature just after the mixing point ($T_{\text{after mix}}$). Due to repeated occurrence of leakage at the connection point of the fine thermocouples with the high pressure tubing, thermocouples of a larger diameter (O.D. = 1.5 mm) built in a 1/8 in. capillary tube via hard-soldering were used for the more explorative experiments: for the study of the influence of the starting material on the synthesis of the iron oxide nanoparticles as reported in Chapter 6 and for the in situ functionalization with carboxylic acids, reported in Chapter 7). All other temperature measuring points were equipped with these thermocouples (type K, O.D. 1.5 mm) from the beginning. However these were built in 1/4 in. high pressure tubings which were then screwed into 1/4 in. union tees or crosses (Swagelok).

Two piezoresistive pressure sensors were used to measure the process pressure (K-Line, absolute pressure, Kistler, Winterthur, Switzerland, precision of calibration for the measurement chain: $\pm$ 0.04 MPa). The mass flow rates of the main water stream and the quench stream—in case of using one of the two double diaphragm metering pumps—were detected with coriolis flow meters (Promass A, Endress + Hauser, Weil am Rhein, Germany, precision of the calibration: 0.01kg/h). A simplified flow diagram showing all the measuring points is displayed in Figure 4.5

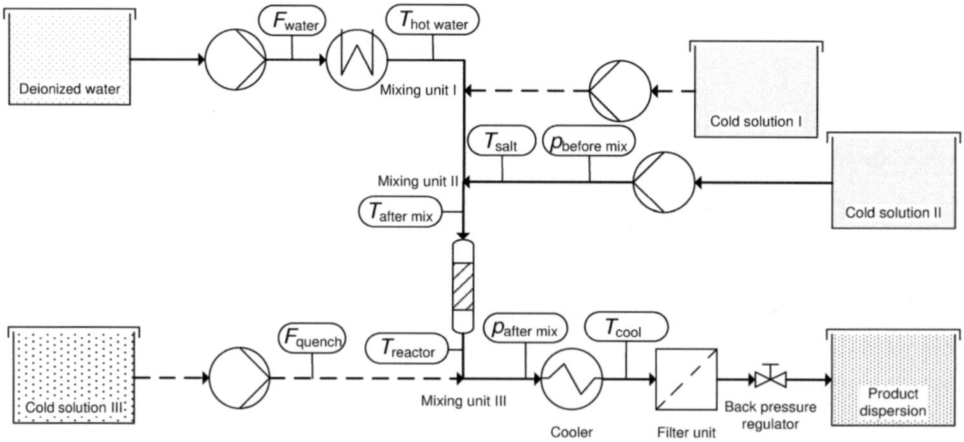


**Figure 4.5:** Simplified flow diagram with the measuring points for the process temperature ($T$), pressure ($p$) and flow rate ($F$).

The analog signals of the temperature indicators, pressure transducers and flow meters were periodically recorded via analog-digital converters (USB-1608FS, Measurement Computing, Plug-in, Eichenau, Germany) using a data logging program (Dasylab, measX, Mönchengladbach, Germany). An output file with the values for the process parameters under consideration of the calibration as a function of process time was generated for each recorded period. Combined with details on the specific conditions, a summary for each recorded period (linked to specific sample of an experimental run) was created using a batch script programmed in R (R Development Core Team, 2011). An example file is shown in Figure C.5 starting on page 214).

Each electrical heating block (with the exception of the one that was placed directly before the first or only mixing unit) was connected to a control unit (custom made with the integration of a PID temperature controller from Eurotherm) setting the pulsation pattern for the heating elements to reach and maintain the set temperature. For the first three electrical heating blocks, the relevant thermocouple was placed at the outside of the wound high pressure tubing at a spot close to the exit of the respective heating block. The remaining electrical heating block was controlled through Dasylab. The process pressure was adjusted by a back pressure regulator (26-1700 Series, Tescom Europe, Selmsdorf, Germany, equipped with Kalrez® seals).

### 4.1.6 Safety

Special attention had to be paid to safety while operating the pilot plant due to the combination of high pressure and high temperature and the synthesis of potentially harmful nanoparticles. Therefore, the entire pilot plant (with the exception of the cold part where the high pressure tubings contained only pressurized water at room temperature) was encapsulated in a housing using MiniTec profile system (MiniTec, Waldmohr, Germany) with removable acrylic glass windows (thickness 6 mm). The top was equipped with a custom made (at Fraunhofer ICT) cover out of stainless steel containing an opening for a flexible hood removing air at a rate of up to 500 $m^3\ h^{-1}$. In case of the occurence of a small leak, all streams were immediately switched to pure water to avoid the release of nanoparticles in form of an aerosol. For the safe operation of the pilot plant, a check list was created to be followed each time the pilot plant was operated (for the check list see Figure C.6 starting on page 216).

## 4.2 Different configurations of the pilot plant

Throughout the work presented in this thesis, the reaction zone always consisted of a 250 mm section of capillary tube (framed by two T-unions for the introduction of the thermocouples) in addition to the reaction volume contained within the mixing unit(s).

For the synthesis of cerium oxide nanoparticles (see Chapter 5), one mixing unit (standard Swagelok 1/8 in. T-union) was in use. For the synthesis of the iron oxide nanoparticle dispersions using different starting materials (see Chapter 6) either one mixing union (1/4 in. Sitec T-union) or two of the same type connected in series as close as possible were used. For the latter case the entry for the cold stream ot the T-union that was further apart from the reaction capillary was closed with a blank plug.

Different mixing setups were implemented for the in situ functionalization of iron oxide nanoparticles with carboxylic acids (see Figure 4.6). In all four configurations, two T-unions as shown in Figure 4.4 were in use. For the setup "salt-organic", the cold aqueous metal salt was combined with the hot compressed water stream before the introduction of the stream containing the carboxylic acids. For the setup "organic-salt", the connection of the two cold streams was reversed. For the setup "mixed", either the two cold streams were mixed by using a T-union before the introduction as a combined stream into the hot compressed water stream or the starting solution contained already the carboxylic acids within the aqueous metal salt solution. In both cases, the second T-union was in use whereas the entry of the first T-union was closed with a blank plug.

A special setup was in use for the functionalization of the iron oxide nanoparticles with 10-undecynoic acid due to the fact that it is a solid at room temperature. Therefore, 10-undecynoic acid was dissolved in ethanol and introduced as an ethanolic stream into the hot compressed stream after the mixing point with the cold aqueous metal salt stream. Furthermore, an ammonia solution was introduced using the quenching stream line to avoid the precipitation of 10-undecynoic acid upon cooling.

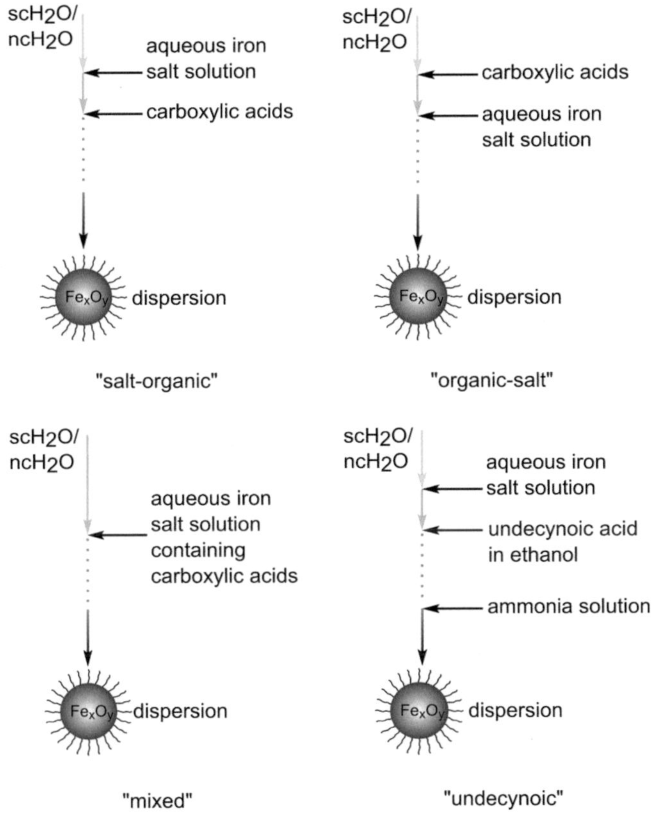


**Figure 4.6:** Mixing setups for the in situ functionalization.

## 4.3 Operation of the pilot plant

The general procedure for the safe startup and shutdown of the pilot plant is described below. This procedure was used throughout the work, also with minor variations depending on the specifics of the experimental run.

1. The deionized water to be used was sparged with nitrogen to reduce the oxygen content and cooling water was started to run through the cooling devices.
2. After sparging, the deionized water was transported into the pilot plant at the required flow rate for the first experimental run to be carried out. Thereafter, the process pressure was slowly increased up to a pressure between 25 and 30 MPa. Once the desired pressure had been reached, the electrical heating blocks were started and the set temperature was chosen such that the hot compressed water reached a temperature at the measuring point $T_{\text{hot water}}$ to give the reaction temperature upon mixing with the cold stream(s). This temperature was initially estimated by applying heat and mass balances.
3. When the hot compressed water had reached the appropriate temperature, the stream line for the aqueous metal salt solution was started to run with deionized water. The set temperatures of the control unit for the electrical heating blocks were fine tuned so that the appropriate temperatures at the measurement points were reached. When a steady state had been reached, the cold stream was switched to the aqueous metal salt solution, which had been sparged with helium prior to its use in order to reduce the oxygen content. Additional cold streams were switched on at this time if required for the specific experimental run.
4. Sufficient time was allowed to pass until the first sample was taken (varying depending on the setup and flow rates for the different experimental runs). For each sample, the development of the measured process parameters was recorded (considering the time it took a fluid segment to flow from the mixing to the sampling point). One or more samples were taken depending on the specific problem being addressed. Thereafter, the process conditions where changed by: i) switching the cold metal salt stream to deionized water, ii) establishing the new process conditions, and iii) switching again to the aqueous metal salt stream.

5. After taking the last samples, all streams where switched to deionized water. Then, all heating elements were switched off. When a temperature below 373 K was reached, either the process pressure was directly reduced to ambient pressure or pressure surges were applied to remove particle deposits from the walls prior to shutdown. The filters were regularly back washed and replaced.

### 4.3.1 In situ functionalization of iron oxide nanoparticles

For the in situ functionalization of the iron oxide nanoparticles as reported in Chapter 7, deionized water was generally delivered into the plant with a flow rate of 80 mL $min^{-1}$ and heated appropriately in order to obtain the desired reaction temperature after mixing with the cold metal salt stream and/or the stream containing the organic compounds. An aqueous iron salt solution with a concentration of 0.0275 mol $L^{-1}$ was delivered with a flow rate of 20 mL $min^{-1}$ into the plant and mixed with the hot compressed water stream. After passing the reaction zone (consisting of the T-unions and high-pressure capillaries), the stream was cooled to a temperature $< 298$ K with an external water jacket. The residence time from the entry into the second mixing unit (for the mixing setups "salt-organic" and "organic-salt", refer to Figure 4.6 for the mixing configurations) or only mixing unit (for mixing setup "mixed") up to the entry into the cooler was from about 300 ms to 600 ms (calculated on the basis of the density of pure water under the process conditions). Thereafter (with the exception of setup "undecynoic", see below), the stream was recovered as a particle suspension after passing through the filter unit and the back pressure regulator installed to control the process pressure of 30 MPa.

For the functionalization of the particles with 5-hexynoic acid at different reaction temperatures and using two different starting materials (iron(III) nitrate nonahydrate and ammonium iron(III) citrate), the particle syntheses were carried out using three mixing variants (4.6 "salt-organic", "organic-salt", "mixed"):

**"salt-organic":** The 5-hexynoic acid was delivered with a flow rate of 0.9 mL $min^{-1}$ into the plant and mixed with the hot stream in a second mixing unit.

**"organic-salt":** The 5-hexynoic acid was delivered with a flow rate of 0.9 mL $min^{-1}$ into the plant and mixed with the hot stream in the first mixing unit before the addition of the cold metal salt stream, which was mixed with the hot compressed water stream in the second mixing unit.

**"mixed":** Always 45 ml of 5-hexynoic acid were mixed with 1 L of the aqueous 0.0275 mol $L^{-1}$ metal salt solution and the organic/salt mixture was delivered with a flow rate of 20.9 mL $min^{-1}$ into the plant and mixed with the hot stream in one mixing unit.

For the functionalization of the iron oxide nanoparticles with 10-undecynoic acid (mixing setup "undecynoic"), the hot compressed water stream was first mixed with the aqueous metal salt stream of 20 mL $min^{-1}$ and thereafter joined with a stream of 15 mL $min^{-1}$ containing 9.62 wt% of 10-undecynoic acid in ethanol. To prevent precipitation of excessive 10-undecynoic acid, the main stream was further mixed directly before entering the cooler with a stream of a 0.82 wt% ammonia solution having a flow rate of 25 mL $min^{-1}$ (Run 7A and 7B, see Table 7.1 in Chapter 7) or 10 mL $min^{-1}$ (Run 7C and 7D in the same table) using a T-type mixing unit (Swagelok).

# Chapter 5

# Synthesis of cerium oxide nanoparticle dispersions

The synthesis of cerium oxide from aqueous cerium nitrate solutions was chosen as a model system (cf. Chapter 1 on page 3) to gain details on the performance of the newly built pilot plant (see Figure 5.1 for the flow configuration in its core part and the geometry of the central mixing unit). The experiments described in this chapter were meant to (i) test the reproducibility and reliability of the process with replicate runs at the same nominal conditions and (ii) test the reliability and stability of the process at different cerium content.

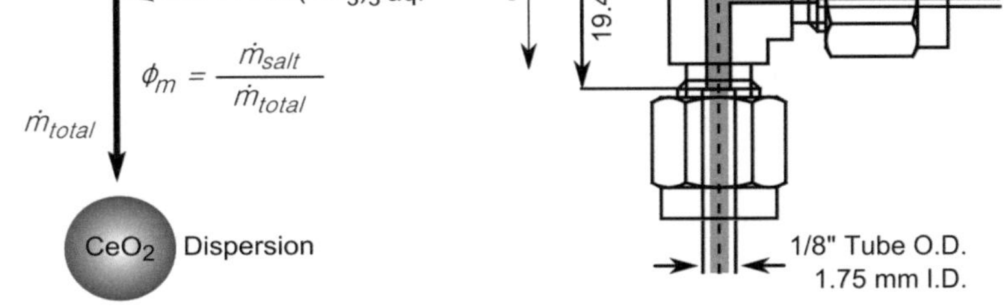


**Figure 5.1:** (left) Flow configuration in the core part of the pilot plant. (right) Geometry of the central mixing unit (standard Swagelok union tee).

## 5.1 Structure, morphology, and dimensions of the synthesized particles

Nanosized crystals were always obtained under the investigated experimental conditions. Furthermore, PXRD analysis revealed that $CeO_2$ of a cubic structure was synthesized in all cases. A representative diffraction diagram is given in Figure 5.2, and the process conditions and corresponding average crystallite size ($d_{crystal}$) determined from the diffraction patterns are shown in Table 5.1.*

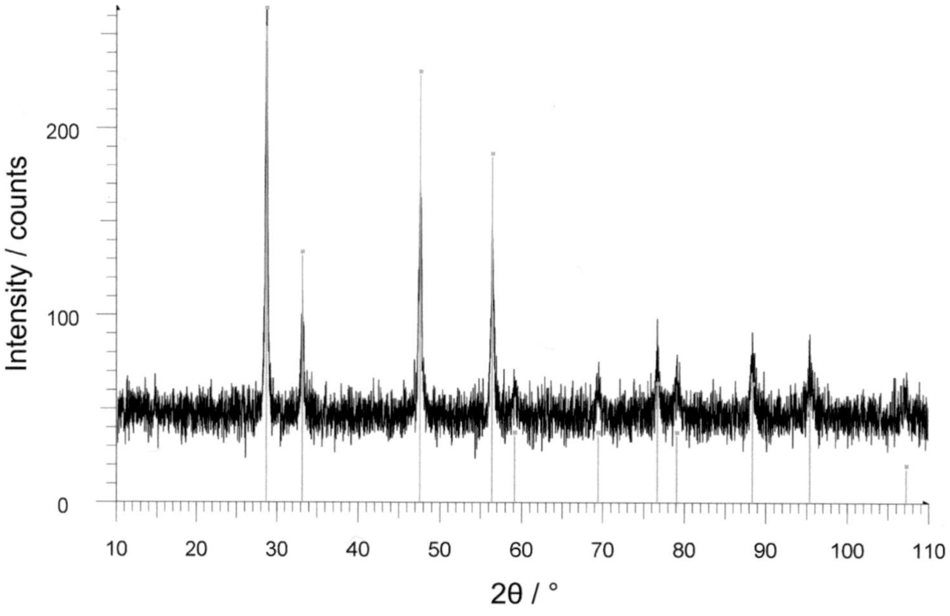


**Figure 5.2:** Representative X-ray diffraction pattern from the $CeO_2$ nanoparticles.

The residence time from the mixing point up to the entry into the cooler was $\approx$ 0.3 s for runs V0-V4 and $\approx$ 0.2 s for runs V5 and V6. Experiments V0-V3 were carried out to explore the reproducibility of setting the process conditions as well as of the properties of the obtained products. This revealed that the variability of the temperature within one experimental run was very small and similar for all runs. The maximum

*The calculation of average crystallite size was performed on the basis of whole pattern fits. For an example see Figure D.1 on page 219.

difference in average temperature between different runs was higher than the variability within runs but always below 2 K. The variability (standard deviation) of the process pressure within a run was $\leq 0.05$ MPa except for run V3 (about three times larger compared to the other runs). The mean deviation between set value (30 MPa) and observed pressure was $\leq 0.33$ MPa.

**Table 5.1:** Process conditions and corresponding mean crystallite size (PXRD) for the experiments performed to synthesize cerium oxide. The uncertainty shown corresponds to one standard deviation of the respective variable throughout the course of the experimental run.

| Run | $c_{salt}$ mol L$^{-1}$ | $\phi_m$ - | $\dot{m}_{total}$ kg h$^{-1}$ | $T_{after\,mix}$ K | $p_{before\,mix}$ MPa | $\Delta p$ MPa | $d_{crystal}$ nm |
|---|---|---|---|---|---|---|---|
| V0 | 0.05 | 1/10 | $6.12 \pm 0.03$ | $674.0 \pm 0.2$ | $30.33 \pm 0.03$ | $-0.02 \pm 0.01$ | 42 |
| V1 | 0.05 | 1/10 | $6.12 \pm 0.03$ | $673.2 \pm 0.2$ | $29.97 \pm 0.05$ | $0.16 \pm 0.04$ | 33 |
| V2 | 0.05 | 1/10 | $6.11 \pm 0.03$ | $672.9 \pm 0.2$ | $30.05 \pm 0.04$ | $0.07 \pm 0.05$ | 44 |
| V3 | 0.05 | 1/10 | $6.12 \pm 0.03$ | $672.1 \pm 0.2$ | $30.14 \pm 0.14$ | $0.02 \pm 0.03$ | 38 |
| V4 | 0.005 | 1/10 | $6.12 \pm 0.03$ | $673.2 \pm 0.2$ | $29.87 \pm 0.03$ | $-0.05 \pm 0.02$ | 31 |
| V5 | 0.05 | 1/15 | $9.16 \pm 0.03$ | $673.2 \pm 0.2$ | $30.18 \pm 0.10$ | $0.30 \pm 0.03$ | 31 |
| V6 | 0.5 | 1/15 | $9.16 \pm 0.04$ | $672.5 \pm 0.3$ | $30.27 \pm 0.29$ | $0.66 \pm 0.08$ | 31 |

The average crystallite size of particles from runs V0, V2 and V3 was 41.3 nm, with deviations of $< 10\%$ between this mean value and the average crystallite size from the individual experimental runs. However, the average crystallite size for run V1 was $\approx 10$ nm smaller than the average of runs V0, V2 and V3. It is important to note that the pressure difference for run V1 was much higher than for V0, V2 and V3. Furthermore, run V1 had to be terminated prematurely due to clogging of the mixing unit. This suggests that the changes in the flow conditions due to deposited particles at the entry of the streams into the mixing unit strongly influence the average crystallite size.

Remarkably, particles with almost the same mean crystallite size (differences $< 1$ nm) were obtained at three different concentrations of the metal salt stream, stretching over three orders of magnitute (runs V4 to V6) despite the variability in process conditions. Whereas the measured fluctuation range of the temperature remained small even up to the highest cerium content, the fluctuations in pressure increased with increasing cerium content (from lowest to highest about one order of magnitute). Parallel to this increasing variability in pressure, the pressure difference ($\Delta p$) measured after and before the mixing and reactor unit also increased significantly. It appears therefore clear that increasing cerium (and consequently particle) content negatively affects process stability. This is likely due to the accumulation and subsequent sudden breakaway of particles from the tube walls due to unfavorable flow conditions especially in the mixing unit.

Particles from runs V3 and V4 were additionally characterized by means of TEM. Representative images are shown in Figure 5.3. The TEM images show individual as well as collective particles were present for run V3. The size of the isolated particles was in the range of 15–50 nm, in which the average crystallite size from PXRD analysis falls. This, together with high resolution imaging, revealed that these particles were mostly single crystallites. An example is shown in Figure 5.3b. Here, the uniform orientation of the atom layers is clearly visible. Particles from run V4 contained a much larger fraction of individual crystallites (see Figure 5.3c). These particles mostly possesed the typical marked forms with sharp edges for single crystals. Comparing image a) with c) suggests that a much larger fraction of the particles were already agglomerated and/or aggregated when obtained as dispersion from the process. However, consolidated findings need to be based on direct measurements of the particle size distribution in suspension. To accomplish this, the particle size distribution of the dispersions of run V1-V6 were determined with high resolution by differential centrifugal sedimentation (DCS).

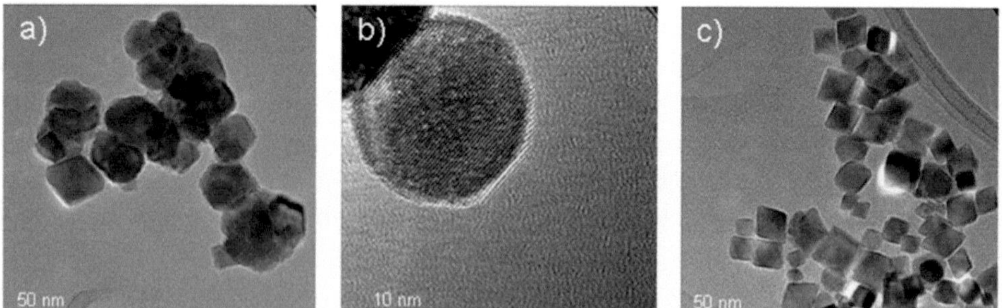


**Figure 5.3:** Transmission electron microscopy images of particles from run V3 (a, b) and V4 (c).

The cumulative particle size distribution ($Q_3$) for the experimental runs V1-V6 is shown in Figure 5.4 and characteristic values of these distributions are listed in Table 5.2. It is obvious that, with increasing product concentration, the particle size distributions—when neglecting the fraction of the biggest 10%—shifted to larger particle sizes. This is congruent with the expectation that more and larger aggregates are formed at higher product concentrations because the probability increases for particle-particle collisions producing coagulated primary particles. For the experimental run V4, the median was 40 nm—in the range of the average crystallite size. In contrast, for experiments carried out with a ten times increased metal salt concentration values of almost 70 up to 85 nm were obtained which is in the order of twice the average crystallite size. The particle size distributions for run V6 (highest metal salt concentration) had a median of around 120 nm. Moreover, the particle size distribution of the experiment with the lowest product concentration (V4) showed the steepest distribution noticeable also at the by far lowest dispersity. The disper-

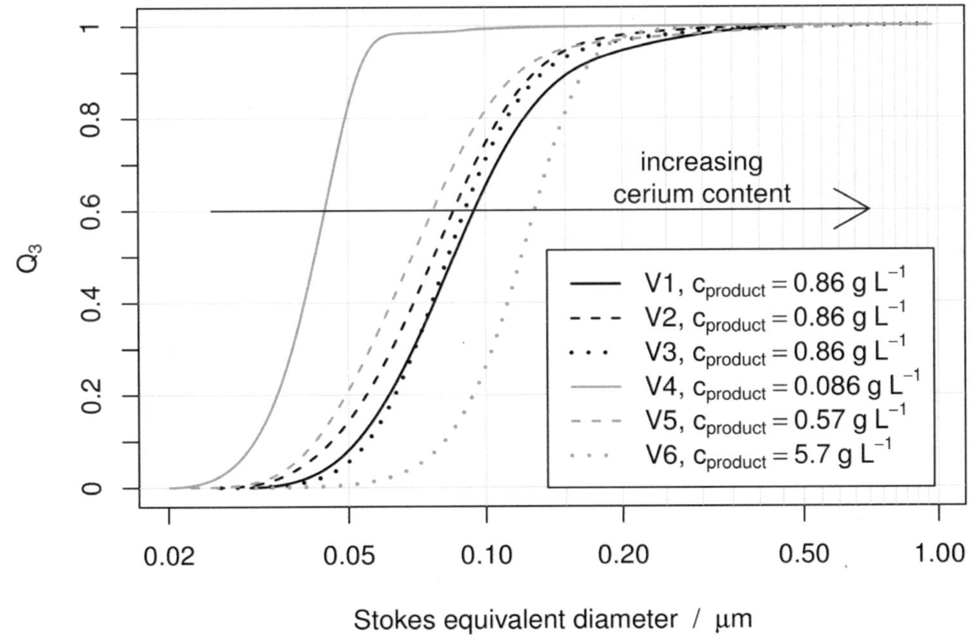


**Figure 5.4:** Cumulative particle size distributions based on mass ($Q_3$) of the dispersions obtained in run V1-V6 (see Table 5.2). The particle size distributions were measured with an analytical disc centrifuge.

sity value for this run falls within the lower quarter of the range classified as "quasi mono-modal". In contrast, the dispersity values of the experimental runs with product concentrations† of one order of magnitude higher fall within the transition region of "quasi mono-modal" to "polydisperse". However, for the experimental run with the highest product concentration the size distribution was again narrower.

**Table 5.2:** Characteristic values of the particle size distributions of run V1-V6 from differential sedimentation analyses. The mean as well as the standard deviation from 2 to 9 independant measurements are given.

| | $c_{product}$ / g L$^{-1}$ | $x_{k,3}$ / nm | $x_{50,3}$ / nm | $x_{16,3}$ / nm | $x_{84,3}$ / nm | $\kappa$ / - |
|---|---|---|---|---|---|---|
| V1 | 0.86 | $83.9 \pm 0.2$ | $85.7 \pm 0.5$ | $58.7 \pm 0.4$ | $132.6 \pm 1.6$ | $0.431 \pm 0.007$ |
| V2 | 0.86 | $87.0 \pm 0.6$ | $77.1 \pm 0.2$ | $51.8 \pm 0.5$ | $115.1 \pm 0.4$ | $0.411 \pm 0.006$ |
| V3 | 0.86 | $80.2 \pm 0.1$ | $82.7 \pm 0.1$ | $60.9 \pm 0.2$ | $119.2 \pm 0.3$ | $0.354 \pm 0.003$ |
| V4 | 0.086 | $42.2 \pm 2.2$ | $40.1 \pm 2.2$ | $32.1 \pm 1.9$ | $48.2 \pm 2.1$ | $0.202 \pm 0.012$ |
| V5 | 0.57 | $68.8 \pm 0.4$ | $69.3 \pm 0.5$ | $47.3 \pm 0.5$ | $104.9 \pm 0.4$ | $0.416 \pm 0.004$ |
| V6 | 5.7 | $124.9 \pm 0.5$ | $119.8 \pm 0.1$ | $89.6 \pm 0.05$ | $154.4 \pm 0.5$ | $0.271 \pm 0.001$ |

Combining the results of the PXRD, DCS and TEM analyses leads to the following picture: increasing the cerium content influences the size of the agglomerates and/or aggregates to a much larger extent than the size of the crystallites. Suspensions with the lowest cerium content contained essentially singly dispersed crystallites whereas the particles in suspensions with the highest cerium content were present nearly only as agglomerates and/or aggregates. Suspensions with intermediate cerium content represent a transition between these two extremes. Here, both singly dispersed crystallites as well as particle agglomerates are present leading to broader distributions and higher values of dispersity.

The mean sizes of the $CeO_2$ crystals fall within the same order of magnitute reported by Hakuta et al. (1998b). However, in this work, the crystals were always larger (from 10 up to 20 nm) under comparable process conditions concerning pressure, temperature and residence time. The fairly high differences in particle size under almost identical process conditions suggests that the described discrepancy may be mainly due to a different flow regime in the central mixing unit. Unfortunately, details on the mixing geometry are not given by Hakuta et al. (1998b).

†This assumes complete conversion, see Section 5.2.

## 5.2 Determination of the conversion

The conversion of the solvated cerium ions into $CeO_2$ was determined by separating the $CeO_2$ nanoparticles out of the product dispersions by membrane filtration and subsequently measuring the cerium and nitrate concentration of the filtrate by ICP-OES and ion chromatography (IC), respectively (cf. Chapter 3). To validate the procedure for the determination of the conversion, test runs were carried out with solutions of a metal salt concentration of the experimental runs V1-V3 ($c_{Salt} = 0.05$ mol/L) using membranes (material PESH) with different molecular weight cut-off (MWCO = 5, 20 and 50 kDa). The cerium and the nitrate concentrations were determined before and after filtration. The results are shown in Table 5.3. In the case of using PESH-membranes with 50 kDa, no significant loss in the cerium as well in the nitrate content was observed in contrast to significant losses in cerium for the two membranes with lower MWCO (5 and 20 kDa). In a next step, a product dispersion (containing $CeO_2$ nanoparticles) of a test run with the same metal salt concentration and mixing value ($c_{salt} = 0.05$ mol/L and $\phi_m = 0.1$) were treated by membrane filtration using membranes with MWCO of 50 kDa. Using this method, a mean conversion of 99.93% (range 99.85-99.98%) was estimated for run V2 and 99.98% for run V3. Because some particles may break through the membrane (thereby increasing the cerium concentration in the filtrate), these values represent a lower bound for the

**Table 5.3:** Results of the determination of the cerium and nitrate concentrations by ICP-OES and IC before and after the treatment by membrane filtration (MF) with membranes (material PESH) of different MWCO. The cerium nitrate ($Ce(NO_3)_3$) concentration of the model solution was 50 mmol/L. The model suspension was obtained by using the model solution as starting material in the hydrothermal process with a mixing value ($\phi_m$) of 0.1.

| | | Concentration in mmol/L $\pm\ \sigma$ | |
|---|---|---|---|
| Sample | MWCO | Nitrate / 3 | Cerium |
| model solution before MF | - | $50.0 \pm 0.5$ | $50.0 \pm 0.5$ |
| model solution after MF | 5 kDa | $47.7 \pm 0.5$ | $46.5 \pm 0.5$ |
| model solution after MF | 20 kDa | $49.0 \pm 0.5$ | $48.8 \pm 0.5$ |
| model solution after MF | 50 kDa | $49.7 \pm 0.5$ | $50.7 \pm 0.5$ |
| model suspension after MF | 50 kDa | $5.0 \pm 0.5$ | $0.2 \pm 0.01$ |

conversion. This is in good agreement with results reported by Hakuta et al. (1998b) (99.9%) using similar process conditions.

## 5.3 Stability of the product dispersions

The product dispersions of initial experiments showed no apparent changes over the course of a few weeks indicating that the dispersions were stable. This warranted a more detailed examination. For this, particle size distributions of the same product dispersion (from experimental run V2) were taken after different days of storage and compared. A superposition of the cumulative distributions is shown in Figure 5.5. A very high degree of consistency is observable. Examination of the characteristic values of these distributions (Table 5.4) reveals marginal changes in particle size with increasing storage time.

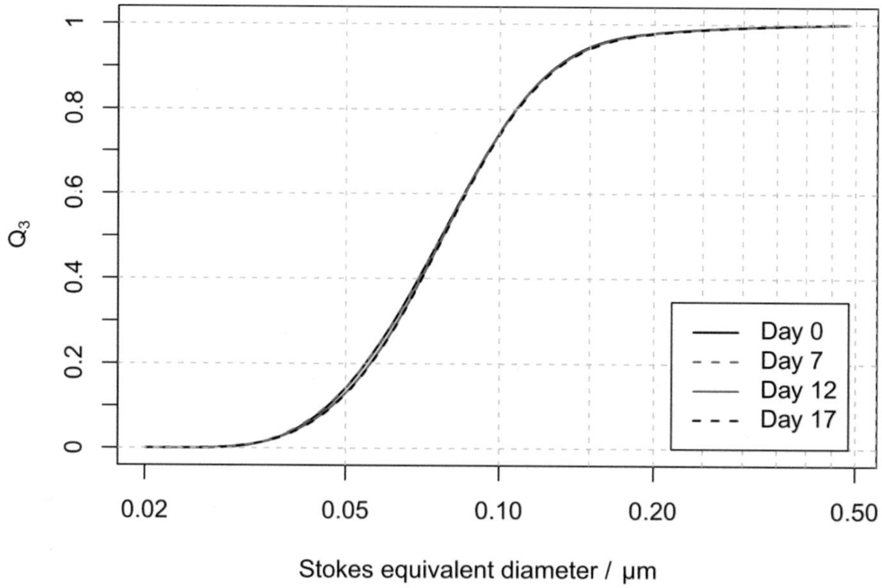


**Figure 5.5:** Cumulative particle size distributions ($Q_3$, measured with an analytical disc centrifuge) for the product dispersion of experimental run V2 measured after different storage times.

**Table 5.4:** Characteristic values of the particle size distributions of run V2 at the day of production and after 7, 12 and 17 days of storage under ambient conditions.

| Days of storage | $x_{50,3}$ / nm | $x_{16,3}$ / nm | $x_{84,3}$ / nm |
|---|---|---|---|
| 0 | 77.1 | 51.8 | 115.1 |
| 7 | 77.0 | 51.7 | 115.1 |
| 12 | 77.6 | 52.8 | 115.2 |
| 17 | 77.8 | 53.1 | 116.0 |

Furthermore, some samples (from runs V3 and V4) were characterized by electrophoretic mobility measurements to ascertain the plausibility of a purely electrostatic stabilisation mechanism. The product suspensions were measured as received with pH-values of 2.2 and 2.8, respectively. Mobilities of $(3.55 \pm 0.08) \times 10^{-8}$ $m^2/Vs$ and $(3.31 \pm 0.17) \times 10^{-8}$ $m^2/Vs$, respectively, were determined. These correspond to zeta potential values of $(45.3 \pm 1.0)$ mV and $(42.3 \pm 2.1)$ mV according to the Smoluchowski model. Because these values are well over 30 mV (a guide value for the electrostatic stability of suspensions), it appears reasonable to assume that the process produces dispersions which are well stabilized by an electrostatic mechanism.

## 5.4 Interim summary

The newly-built high pressure pilot plant described in Chapter 4 was used to produce electrostatically stabilized cerium oxide nanoparticle dispersions in a continuous hydrothermal process with cerium nitrate solutions as starting material and conversions $> 99.8\%$ with residence times in the reactor $< 1$ s. The cerium oxide nanoparticles had mean crystallite sizes in the range of 30-45 nm. These crystallites were shown to be mostly singly dispersed at low cerium content and to form agglomerates and/or aggregates with increasing cerium content. However, the size of the cerium oxide crystallites did not show a dependence on the cerium content of the suspensions (varied over three orders of magnitude).

The relatively poor reproducibility concerning the mean crystallite size observed in the replicate runs and sporadical clogging of the mixing unit point to the need for optimizing the mixing geometry.

# Chapter 6

# Synthesis of iron oxide nanoparticle dispersions

In due consideration of the experience with the newly-built high pressure pilot plant with the model system cerium oxide (cf. Chapter 5), the continuous hydrothermal method was used to prepare iron oxide nanoparticle dispersions without the addition of organic solvents, strong bases or surfactants. Of the many forms of iron oxide, hematite, magnetite and ferrihydrite nanoparticles have been synthesized successfully in hot compressed water (Lam et al., 2008). However, both structure and properties of iron oxide nanoparticles strongly depend on their particle size, their history of formation and the surrounding medium (Chernyshova et al., 2007). Thus, the goals of this study—addressing the relationship between process conditions and product properties—were twofold:

1. to investigate the influence of mixing ratio, fluid velocity and identity of the starting material on the size distribution of both primary particles and particle agglomerates, as well as on the colloidal stability of the dispersions, and
2. to identify the structure of the particles directly in the aqueous dispersion by means of Raman spectroscopy. This second goal is important because the properties of the iron oxides can be changed during preparation steps for characterization as a powder.

To accomplish the first goal, on the one hand the mixing value ($\phi_m$) and the total mass flow rate ($\dot{m}_{total}$) were systematically varied keeping the iron content in the product dispersion constant, and on the other hand the ratio between two different iron salts (iron(III) nitrate and iron(II) acetate) and the addition of nitric and/or acetic acid was systematically varied under favorable flow conditions (see Figure 6.1). Due to the problems encountered with the standard mixing T-unit from Swagelok, a mixing unit from Sitec-Sieber with custom made transitions to the adjacent tubings was used. In contrast to the Swagelok component with varying cross-section areas, the new mixing unit had an inner diameter of 1.6 mm throughout. Therefore, abrupt changes in the cross-sectional area in the mixing unit which cause unfavorable flow conditions with the high risk of clogging could be avoided (for more details on the mixing cf. 4.1.3). A summary of the process conditions is shown in Table 6.9.

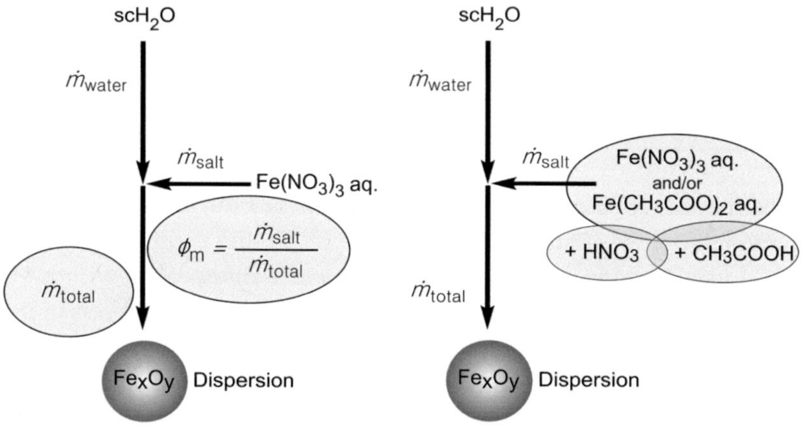


**Figure 6.1:** Schemes of the flow configuration in the core part of the pilot plant; (left) for the study on varying the flow conditions; (right) for the study on varying the chemical compositions of the metal salt stream.

**Table 6.1:** Summary of process conditions.

| Parameter | $p_{after\,mix}$ MPa | $T_{after\,mix}$ K | $\phi_m$ - | $\dot{m}_{total}$ kg h$^{-1}$ | $c_{\text{product Fe}}$ mol L$^{-1}$ |
|---|---|---|---|---|---|
| set value | 30 | 673 | 0.05 - 0.3 | 3.04 - 9.11 | 0.001 - 0.01 |

To accomplish the second goal, a new measurement setup for performing meaningful raman spectroscopic investigations of the iron oxide nanoparticle dispersions had do be developed. This setup was designed by Prof. Gerd Irmer and his group at the Institute of Theoretical Physics of the TU Bergakademie Freiberg. Details are given in Section 3.2.3.

# 6.1 Dispersions synthesized under different flow conditions

In order to investigate the effect of different flow conditions on the product properties and their reproducibility, a series of experiments were carried out using iron(III) nitrate as the starting material ($c_{\text{product Fe}} = 0.0055$ mol/L). The resulting particle dispersions exhibited an intense color (mostly red) but both the color and the degree of transparency varied depending on the process conditions, as shown for a selection of samples in Figure 6.2. It is immediately apparent that an increase in transparency—indicating the presence of smaller particles—was accompanied by a shift in color from orange-red (characteristic of suspensions containing hematite particles $< 0.1$ µm, see Figure 6.2a) to more intense reddish colors with a touch of purple (e.g., Figure 6.2c). However, suspensions of hematite particles are known to change their color from orange to red to even purple upon increasing particle sizes (Cornell and Schwertmann, 2003). Alternatively, the change in color may result from oriented aggregates of nanocrystallites, as has been reported for hematite particles synthesized via a different route (Chernyshova et al., 2007; Cornell and Schwertmann, 2003). Finally, the observed color changes may be introduced through size dependent changes in the structure of the particles. This possibility was explored in more detail by means of Raman spectroscopy, as discussed below.

## 6.1.1 Structure of the suspended particles

Due to possible structural changes in the iron oxide nanoparticles upon changes in their environment and especially upon drying, a method was chosen that allows structural investigations of the nanoparticles in the aqueous medium in which they are obtained. On the basis of the work by Chernyshova et al. (2007), Raman spectroscopic investigations of the dispersions seemed the most promising method for this purpose. Therefore, a modified measurement setup was developed in which the sample is examined as a 'hanging drop' in an enclosure (for details see Section 3.2.3). It is known that significant differences in the Raman spectra can be observed when using different excitation wavelengths (Wang et al., 2005). In this study, Raman scattering was excited with the 532 nm line of a frequency-doubled Nd:YAG laser instead of using a 632.8 nm He-Ne laser. For comparison, hematite nanoparticles prepared according to Schwertmann and Cornell (2000) were excited using the 647 nm line of

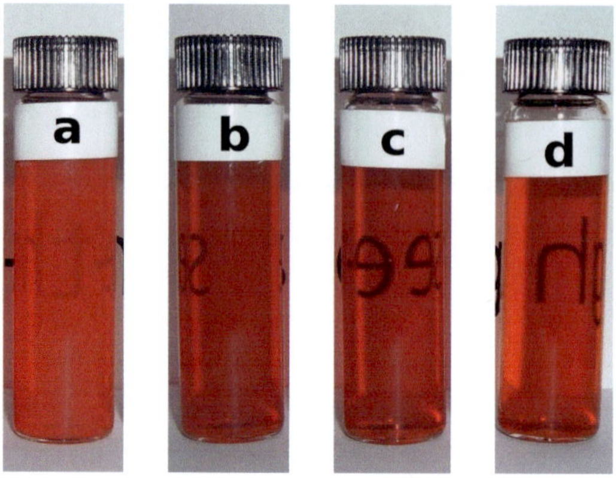


**Figure 6.2:** Selection of dispersions synthesized under different conditions: samples (a) - (c) were synthesized from iron nitrate with $c_{\text{product Fe}} = 0.0055$ mol/L with increasing total mass flow rate ($\dot{m}_{\text{total}}$) from (a) 3.04 kg/h, (b) 6.07 kg/h and (c) 9.11 kg/h, at a constant mixing value ($\phi_m$) of 0.2. Sample (d) was synthesized from acidified ($c_{\text{HNO}_3} = 0.002$ mol/L) iron acetate solution with $c_{\text{product Fe}} = 0.001$ mol/L, $\phi_m = 0.2$ and $\dot{m}_{\text{total}} = 3.04$ kg/h. The photo for (d) was taken after increasing its concentration to give a similar solid matter content as for the samples (a) - (c).

a $Kr^+$ laser in addition. The spectra are shown in Figure E.1 on page 221. With respect to the intense $A_{1g}$ mode at 224 $cm^{-1}$ the spectra were normalized. Exciting with 532 nm the intensity of the mode at 1320 $cm^{-1}$ was strongly enhanced in comparison with the 647 nm excitation. This finding is in agreement with the report by Wang et al. (2005) but the behavior upon wavelength change is still unrevealed. For the structural investigation of the hydrothermally synthesized iron oxide nanoparticles considered in this work, Raman spectra of relevant reference samples were obtained with the same equipment. These are shown in Figure 6.3.

Whereas the position and assignment of hematite single crystals (at about 226 ($A_{1g}$), 245 ($E_g$), 293 ($E_g$), 298 ($E_g$), 412 ($E_g$), 500 ($A_{1g}$) and 612 ($E_g$) $cm^{-1}$; Chernyshova et al., 2007) is well established, effects of the degree of crystallinity, defects, and the particle size are still under discussion.The assignment of the strong band at around 1320 $cm^{-1}$ is controversially debated. However, the dependence of its intensity on the excitation wavelength suggests the scattering by second order processes, either due to a two-phonon or two-magnon scattering process (Lübbe et al., 2010). The first order mode

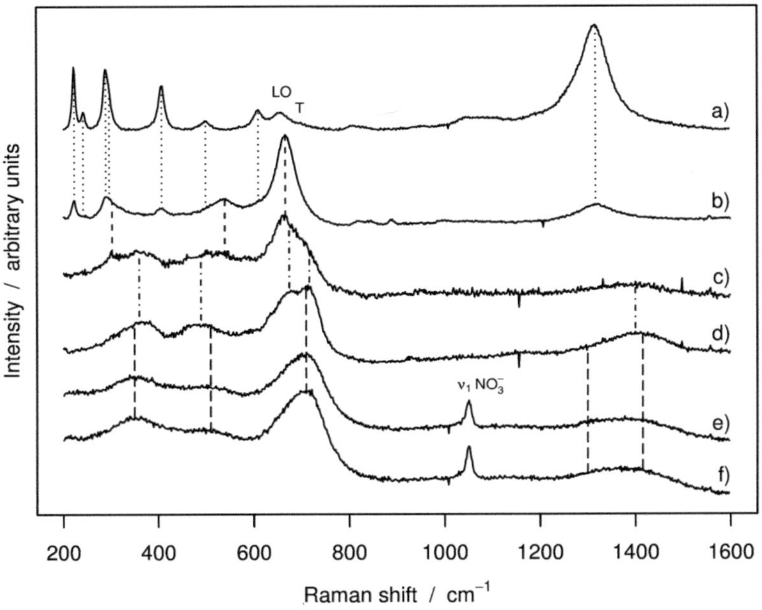


**Figure 6.3:** Raman spectra of: a) freeze dried hematite nanoparticles prepared according to Schwertmann and Cornell (2000); b) natural magnetite; c) iron(II,III) oxide powder (<5 μm, 95%, Aldrich); d) iron(II,III) oxide nanopowder (<50 nm particle size (TEM), ≥98% trace metal basis); e) 2-line ferrihydrite prepared according to Schwertmann and Cornell (2000); f) 6-line ferrihydrite prepared according to Schwertmann and Cornell (2000). The vertical lines between a) and b) correspond to hematite, between b) and c) to magnetite, between c) and d) to maghemite and between d) to f) to ferrihydrite. The spectra are shifted vertically for clarity.

would be the longitudinal optical one at 660 $cm^{-1}$ which is Raman forbidden owing to the $E_u$ symmetry (Lübbe et al., 2010; Wang et al., 2005).

In the rigorous study of Chernyshova et al. (2007) on size and environment–dependent structural transformations of hematite nanoparticles, it was found that two additional lines at about 690 $cm^{-1}$ and 660 $cm^{-1}$ appear for hematite nanoparticles and become more pronounced for smaller sizes. The mode at around 690 $cm^{-1}$ was attributed to tetrahedral defects (T). The band at 660 $cm^{-1}$ has been interpreted differently by various authors. In the more recent studies, the interpretation that this line is due to the Raman forbidden longitudinal optical (LO) $E_u$ mode, which is activated by lattice disorders within the hematite crystal, has proved to be more reasonable (Chernyshova et al., 2007; Jubb and Allen, 2010) and is also adopted for this work. Furthermore, it was found that the T band increases in intensity upon drying of the nanoparticles. This was explained by the creation of tetrahedal sites to counterbalance the increase in the surface energy due to dehydration (or reduction in size of particles). The increase of structural disorder of the hematite nanocrystals and the formation of T defects was seen in the context of the destabilization of the hematite and the size-induced $\alpha$ to $\gamma$ phase transition (Chernyshova et al., 2007). Within that study, the onset of the $\alpha$ to $\gamma$ phase transition for the dry particles was estimated to be around a mean crystallite size of 40 nm based on PXRD-analysis. For the reference hematite nanoparticles sample (see Figure 6.3a), all characteristic lines are observed as well as the two additional lines in agreement with the study of Chernyshova et al. (2007) for structurally destabilized nanoparticles. The diffraction pattern of this and the other reference samples except for the natural magnetite are displayed in Figure E.2 on page 222. In comparison to hematite, magnetite and maghemite are weak Raman scatterers (Jubb and Allen, 2010; Thibeau et al., 1978). Because of this and the fact that they can be easily transformed upon laser irradiation, there are still some controversial issues with the location and assignment of the characteristic Raman bands. However, in contrast to X-ray diffraction where these two phases show almost identical diffraction patterns, Raman spectroscopy allows the discrimination between these phases also in the nanoscale regime (El Mendili et al., 2011; Vangelista et al., 2012). Although for magnetite often only three of the Raman active phonon modes are observed in the frequency range from 200 to 1000 $cm^{-1}$, a fairly strong and sharp band at around 670 $cm^{-1}$ can be defined as diagnostic for magnetite (Hanesch, 2009). This band is clearly visible in the spectrum of natural magnetite (which contains a small fraction of hematite) shown in Figure 6.3b besides two very weak bands at around 540 and 305 $cm^{-1}$, which can also be attributed to magnetite (Hanesch, 2009). In contrast, as in the more recent studies mostly agreed on, Raman spectra of maghemite consist of three broad bands in the low frequency range, namely at around 350, 510, and 700 $cm^{-1}$. The last band often appears as a superposition of two lines, as observable in Figure 6.3d (Hanesch, 2009). Concerning characteristic Raman bands in the high frequency region a single band at around 1400 $cm^{-1}$, as observed for the sample studied in this work, or two bands at around 1400 and 1570 $cm^{-1}$ (e.g. in Jubb and Allen, 2010, where both variants were observed) have been proposed (Chourpa et al., 2005; Mazzetti and Thistlethwaite, 2002). As the band at around 1570 $cm^{-1}$ is not always observable and appears also in Raman spectra of other minerals, in the more recent studies it has not been marked as being characteristic for maghemite in contrast to the band at around 1400 $cm^{-1}$ (Chourpa et al., 2005; El Mendili et al., 2011). An explanation for the sometimes observable band around 1570 $cm^{-1}$ is the presence of burnt organic material (Hanesch, 2009). The purchased iron(II,III) oxide powder of which the Raman spectra is shown in Figure 6.3c turned out to be a mixture of maghemite and magnetite as clearly visible by looking at the shape of the broad Raman band at around 700 $cm^{-1}$ and the very weak but detectable band at around 1400 $cm^{-1}$, supporting the possibility of discrimination between maghemite and magnetite and the localization of the characteristic Raman bands.

Similarly, some controversy still exists on the characteristic Raman bands of ferrihydrite. Nevertheless, it appears most reasonable to identify three broad bands at around 370, 510 and 710 $cm^{-1}$—this last band is much stronger than the other two ones and very asymmetric in shape—as being characteristic for ferrihydrite (Das and Hendry, 2011; Hanesch, 2009; Mazzetti and Thistlethwaite, 2002). These bands were also observed for both ferrihydrite reference samples investigated within this work (see Figure 6.3e and 6.3f). However, the band at around 370 $cm^{-1}$ appeared always slightly lower (at around 350 $cm^{-1}$) than reported in the literature. This deviation appears minor considering the breadth and low relative intensity of this band. Furthermore, we consistently found a broad band, which might be a superposition of two bands, in the high frequency range from around to 1300 to 1400 $cm^{-1}$. This was also observable in the studies of Hanesch (2009) and Mazzetti and Thistlethwaite (2002). The only difference we observed between the so called 2-line- (in Figure 6.3e) and 6-line-ferrihydrite (6.3f), was that the strongest band of the 6-line sample was not as asymmetric as the one from the 2-line-ferrihydrite. This, in combination with the PXRD patterns (shown in Figure E.2 on page 222) could be an indication of the presence of a small fraction of maghemite in the sample into which ferrihydrite can transform (Mazzetti and Thistlethwaite, 2002). Here, it is important to note that a fairly strong similarity exists between the Raman spectra of maghemite and ferrihydrite samples. However, the difference in the shape of the strongest band at around 700 $cm^{-1}$ seems sufficient allowing a discrimination of these two iron (hydr)oxide species. In addition, the spectral position of the high frequency band is also shifted analyzing maghemite or ferrihydrite. Furthermore, it is important to mention that the structure of ferrihydrite is still debated (Michel et al., 2007; Peak and Regier, 2012). Especially, a very recent study gives strong evidence for the existence of quite a substantial fraction of tetrahedrally coordinated Fe(III) in the structure of ferrihydrite

(Peak and Regier, 2012). Within that study, a strong similarity between maghemite and ferrihydrite was also found by directly probing the local electronic structure using Fe L-edge XANES. The band around 690 $cm^{-1}$ is related to tetrahedrally coordinated iron(III) as already discussed for hematite nanoparticles containing maghemite-like (meaning tetrahedral) defects. Peak and Regier (2012) stated that the formation of under-coordinated Fe(III) at the surface of dried samples upon freeze-drying is a possible source of tetrahedral Fe(III) in ferrihydrite. The freeze-drying process is commonly used for the preparation of minerals. However, detailed structural investigations of ferrihydrite dispersed in water were not found in the literature.

For the characterization of the hydrothermally synthesized particles within this study, the observed values for the characteristic Raman bands for hematite, magnetite, maghemite and ferrihydrite are summarized and compared to values as reported in the more recent literature in Table 6.2 (Das and Hendry, 2011; Hanesch, 2009; Jacintho et al., 2007; Jubb and Allen, 2010; Lübbe et al., 2010; Mazzetti and Thistlethwaite, 2002; Schwertmann and Cornell, 2000; Sousa et al., 2000). Please refer to these and related studies for further details, assignments, and discussion of the Raman active phonon modes.

**Table 6.2:** Raman bands observed for hematite, magnetite, maghemite, and ferrihydrite from the reference substances studied in this work. For comparison, the range of values for the corresponding bands—taken from more recent literature—is displayed next to the value (Chernyshova et al., 2007; Das and Hendry, 2011; Hanesch, 2009; Jacintho et al., 2007; Jubb and Allen, 2010; Lübbe et al., 2010; Mazzetti and Thistlethwaite, 2002; Sousa et al., 2000). Broad bands in the higher frequency range have been observed for maghemite and ferrihydrite.

| hematite | | magnetite | | maghemite | | ferrihydrite | |
|---|---|---|---|---|---|---|---|
| obs. | range (lit.) | obs. | range (lit.) | obs. | range (lit.) | obs. | range (lit.) |
| **224** | 222-229 | 304 | 306-310 | 360 | 350-365 | 350 | 361-370 |
| 243 | 230-249 | 540 | 538-554 | 490 | 500-512 | 510 | 508-510 |
| **291** | 290-295 | **667** | 662-672 | **675** | 655-676 | **710** | 707-710 |
| 298 | 298-302 | | | **717** | 715-730 | | |
| **408** | 408-414 | | | | | | |
| 500 | 490-500 | | | | | | |
| 610 | 607-615 | | | | | | |
| **1317** | 1314-1321 | | | | | | |

Using the 'hanging drop' setup, Raman spectra of several concentrated product dispersions were recorded. Furthermore, some of these dispersions were allowed to air-dry and were subsequently measured as a solid matter in order to compare between dispersed and air-dried particles as well as between hydrothermally prepared particles and the reference samples, which were measured as a solid matter. An overlay of representative spectra is shown in Figure 6.4. Extended Raman spectra in the range from 200 to 4000 $cm^{-1}$ are displayed in Figure E.3 on page 225 for selected samples. Here, the characteristic modes of water as dispersing agent are observable. In Figure 6.4 the Raman spectra of dispersions prepared with two different solid contents under otherwise identical conditions are shown. Besides the expected deviations between the relative intensities of the main water band(s) and the modes due to hematite, no significant differences were observable. The characteristic bands of hematite single crystals (at about 225, 245, 291, 298, 411, 499 and 612 $cm^{-1}$; see discussion on hematite above) are clearly present in all spectra (Figure 6.4) of the samples synthesized using iron(III) nitrate. However, in dependence on the investigated sample the spectra show (i) slight differences in the full width at half maximum (FWHM) of the characteristic hematite modes and (ii) additional bands depicted by the dashed lines in Figure 6.4.

For the synthesized dispersions $c_{\text{product Fe}} = 0.0055$ mol/L, a broad band from around 650 to 750 $cm^{-1}$ is observable in Figure 6.4a and 6.4b which might be a superposition of three modes, namely at about 660 , 690 and 715 $cm^{-1}$. The first frequency is most likely due to the Raman forbidden LO phonon with $E_u$ symmetry which is activated by lattice disorders within the hematite crystal as discussed above. The second band is barely visible in the spectra and can be attributed to previously mentioned tetrahedral defects (T). However, the asymmetric band at about 715 $cm^{-1}$ which is slightly shifted to higher wavenumbers compared to the investigated reference samples (Figure 6.3) as well as the very weak and broad band at about 355 $cm^{-1}$ indicate the presence of particles with maghemite and/or ferrihydrite structure next to structurally disordered hematite nanoparticles. Nevertheless, a discrimination between maghemite and ferrihydrite is not possible in that case as the strong high frequency mode of hematite at 1320 $cm^{-1}$ dominates this spectral range where also bands of ferrihydrite and maghemite appear. Furthermore, as the Raman cross section of hematite is larger than that one of maghemite or ferrihydrite, the relevant signals were caused by a quite substantial fraction of maghemite and/or ferrihydrite. However, more investigations are necessary to support or dismiss this interpretation.

For the dispersions synthesized with the lower $c_{\text{product Fe}}$ content (Figure 6.4c), the relative intensity of the broad band from around 650 to 750 $cm^{-1}$ is much lower than in the above case indicating a smaller fraction of destabilized hematite and/or maghemite/ferrihydrite nanoparticles. However, the spectral position and the width of the hematite bands are very similar pointing to comparable particle sizes.

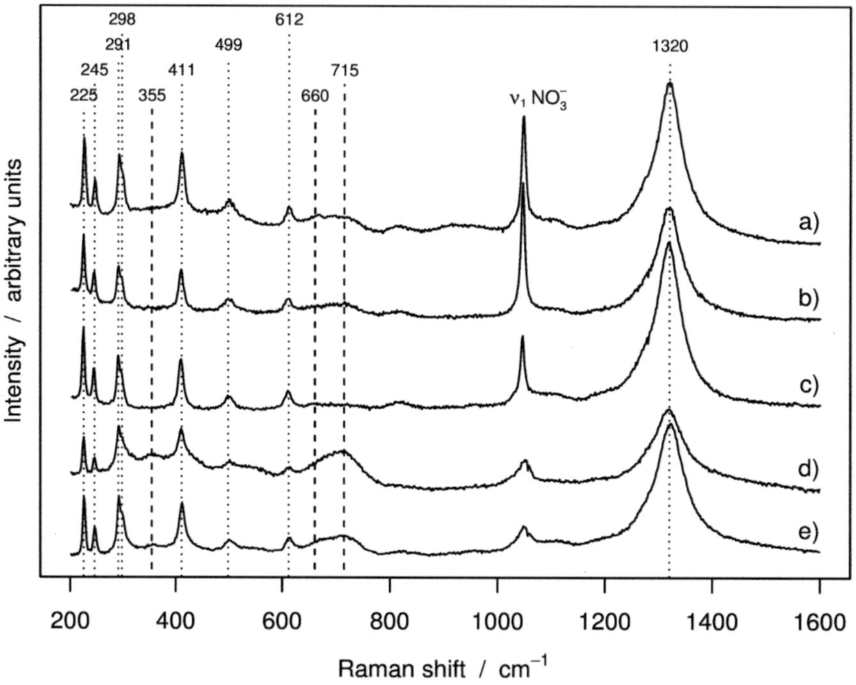


**Figure 6.4:** Raman spectra of iron oxide nanoparticles synthesized using iron(III) nitrate at $\phi_m = 0.2$ and $\dot{m}_{total} = 6.07$ kg/h: a) with $c_{\text{product Fe}} = 0.0055$ mol/L, measured as particle dispersion after increasing the iron oxide content via evaporation of water under vacuum up to about 85 g/L; b) from an independent experimental run carried out under the same condtions as for sample a) but measured as a wet precipitate; c) using iron(III) nitrate with $c_{\text{product Fe}} = 0.001$ mol/L; d) sample b) measured after air-drying; e) sample c) measured after air-drying. The spectra are shifted vertically for clarity.

In addition, droplets of the aqueous suspensions investigated were dried in ambient atmosphere. The obtained Raman spectra of the air-dried samples are displayed in Figure 6.4d and 6.4e. For comparison, the corresponding spectra of the dispersed state are shown in Figure 6.4b and 6.4c, respectively. It is immediately striking that for both samples the spectra changed in shape and relative intensity with respect to the hematite modes in the region from 650 to 750 $cm^{-1}$. Furthermore, the band at about 355 $cm^{-1}$ is clearly visible. Moreover, the hematite bands are broadened in comparison with the reference sample which points to smaller hematite particles. Obviously, the air-drying led to a reduction of the size of the hematite domains. This might be explained by a reduction in the net structural order as well as the introduction of tetrahedral defects (T) to counterbalance an increase in the surface energy upon dehydration. The increase of the LO and T lines compared to the hematite mode at 612 $cm^{-1}$ upon air-drying is an indication for this structural change, as observed by Chernyshova et al. (2007). Furthermore, the band at about 355 $cm^{-1}$ is likely due to ferrihydrite and/or maghemite. This seems reasonable as ferrihydrite is known to be enriched in tetrahedral defects (Jambor and Dutrizac, 1998).

### 6.1.2 Stability of the dispersions

The transparency of the product dispersions and the absence of changes in color—which is very sensitive to any changes in particle and/or agglomerate size for iron oxide suspensions (Cornell and Schwertmann, 2003)—over the course of a few weeks indicated that the dispersions were stable under ambient conditions. Furthermore, the electrophoretic mobility of several samples obtained under different process conditions was measured as $(3.75 \pm 0.14) \times 10^{-8}$ $m^2/Vs$. The $\zeta$ potential derived by applying the model of Smoluchowski and of Hückel was $(47.8 \pm 1.8)$ mV and $(71.7 \pm 2.7)$ mV, respectively. No significant differences could be noticed between the samples. Values of the $\zeta$ potential well above 30 mV stress that the particles were electrostatically stabilized in suspension.

Additionally, the stability of the iron oxide nanoparticle dispersions was investigated upon the addition of different amounts of NaCl as background electrolyte. This is important because the background electrolyte affects the thickness of the electrical double layer of the charged particle surfaces and therefore the stability of the suspended particles. As shown in Table 6.3 NaCl concentrations up to 80 mmol/L could be present without significantly affecting the stability of the suspended particles. As expected for particles being purely stabilized by an electrostatic mechanism, the particles coagulate and sediment quickly beyond a certain concentration of indifferent electrolytes. Partial sedimentation was observed at a NaCl concentration of 200 mM. At concentrations of 400 mM essentially all particles coagulated and sedimented (see Figure 6.5).

**Table 6.3:** Electrophoretic mobility and $\zeta$ potential derived by applying the model of Smoluchowski upon the additon of different amounts of NaCl.

| $c$(NaCl) mM | Conductivity mS/cm | Electrophoretic mobility $\times 10^{-8}$ $m^2/Vs$ | $\zeta$ potential mV |
|---|---|---|---|
| - | $6.64 \pm 0.05$ | $3.80 \pm 0.12$ | $48.4 \pm 1.5$ |
| 20 | $8.88 \pm 0.10$ | $3.70 \pm 0.10$ | $47.2 \pm 1.3$ |
| 40 | $11.1 \pm 0.1$ | $3.60 \pm 0.09$ | $45.8 \pm 1.1$ |
| 60 | $13.4 \pm 0.2$ | $3.76 \pm 0.10$ | $47.9 \pm 1.3$ |
| 80 | $15.6 \pm 0.3$ | $3.54 \pm 0.07$ | $45.1 \pm 0.9$ |
| 100 | $17.9 \pm 0.3$ | $3.29 \pm 0.11$ | $41.9 \pm 1.4$ |
| 200 | $30.0 \pm 0.9$ | $2.83 \pm 0.10$ | $36.1 \pm 1.2$ |
| 400 | $49.3 \pm 0.1$ | $2.23 \pm 0.10$ | $28.4 \pm 1.3$ |
| 800 | $87.5 \pm 0.4$ | $1.56 \pm 0.08$ | $19.9 \pm 1.0$ |

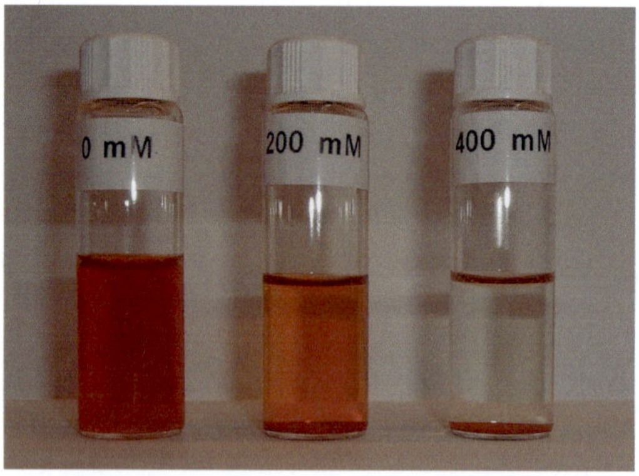


**Figure 6.5:** Dispersions after the addition of different amounts of NaCl.

### 6.1.3 Particle size distributions

The influence of the flow conditions in the mixing unit on the particle formation, with related instabilities in the process conditions which can lead to fluctuations in product properties, was discussed in Chapter 5 and implicitly or explicitly addressed in several publications of other research groups (e.g Blood et al. (2004) and refer to 2.1.2). However, detailed investigations of this effect on the dispersion properties are lacking. Therefore, based on the promising results on the determination of particle size distribution of the cerium oxide nanoparticle dispersions (5) with high resolution and reproducibility, the particle size distributions of the iron oxide nanoparticle dispersions under different flow conditions were measured and compared using differential sedimentation analysis in a disc centrifuge (refer to Chapter 3.2.1). In Figure 6.6, the particle size distributions (shown as $Q_3$ and $q_3$ distributions) of dispersions synthesized at different total mass flow rates ($\dot{m}_{total}$) at two different mixing values ($\phi_m$ = 0.1 and 0.2) are displayed. The particle size distributions appear to be dominated by the superposition of two peaks: for all distributions, the mode of the peak containing the particles with the smaller diameter was around 20 nm in comparison to values from 50 up to 80 nm for the mode of the fraction of particles with larger sizes. However, it is important to note that the mode of the peak containing the smaller particles shifts to lower values of the equivalent spherical diameter with increasing $\dot{m}_{total}$, whereas the opposite effect was mostly observed for the other peak. Furthermore, the ratio between these two particle fractions is highly dependent on the flow conditions set.

Before proceeding with a more in depth examination, it is worth to mention that the material properties of the particles are not exactly known. Therefore, systematic errors concerning the weighting of the two fractions are likely and discussed in 3.2.1. However, the existence of two particle fractions of different sizes indicates that the nuclei of the two size fractions were formed at different locations or that the formation followed a different nucleation and/or growth mechanisms. A reason for that could be inhomogeneous equilibrium conditions during the heating phase of the cold metal salt stream entering the mixer, resulting in the formation of particles of different sizes or different composition. The possibility of particle formation already in the tubing adjacent to the mixer through which the metal salt stream was delivered was noticed in investigations of a different research group—however at much lower flow rates (Blood et al., 2004). In addition, the particles produced in this process are thought to arise from both homogeneous and heterogeneous nucleation, leading to skewed or

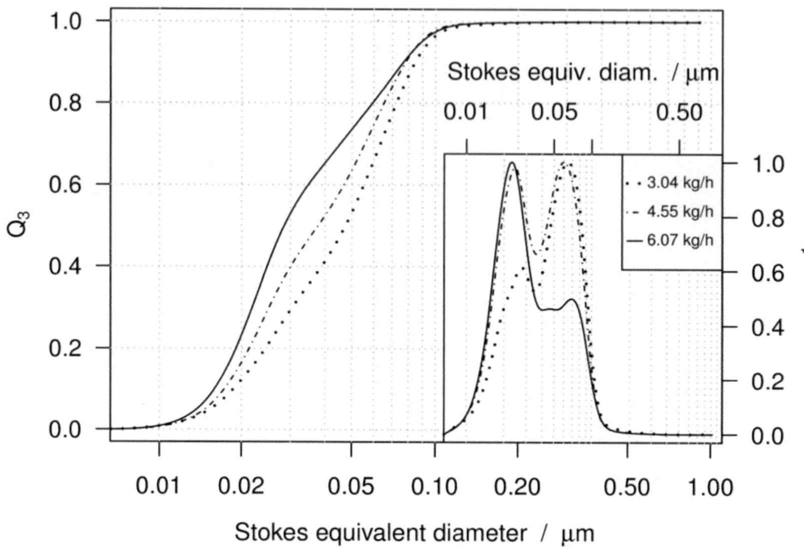


**(a)** $\phi_m = 0.1$

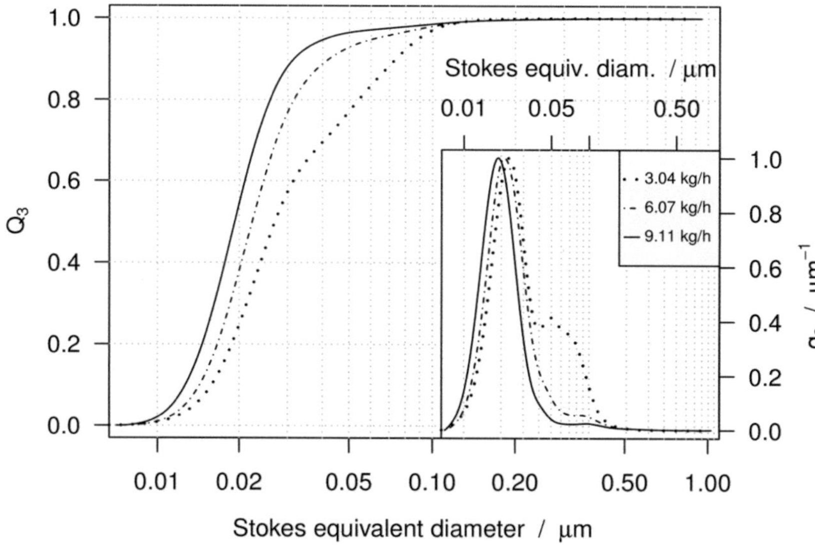


**(b)** $\phi_m = 0.2$

**Figure 6.6:** Particle size distribution ($Q_3$ and $q_3$) of dispersions synthesized at different total mass flow rates ($\dot{m}_{total}$) and mixing values ($\phi_m$).

bimodal particle size distributions. Sue et al. (2010) have recently shown that optimization of the mixer geometry leads to narrower primary particle size distributions (as measured by TEM) and attribute this to the suppression of heterogeneous nucleation for the same material system under identical process conditions (temperature, pressure and mixing ratio) used here. In order to quantify the influence of varying process conditions on the size distribution of the dispersed particles, we used the geometric mean ($x_{g,3}$) and the geometric standard deviation ($\sigma_{g,3}$) calculated for the distributions from the systematic study on varying the flow conditions. These values were plotted against the corresponding values of $\dot{m}_{total}$ and $\phi_m$ and are shown in Figure 6.7.

**Table 6.4:** Characteristic values of the particle size distributions (measured with a disc centrifuge) obtained for dispersions synthesized under different flow conditions. The mean and standard deviations reported were calculated from at least three measurements of the same sample. The values displayed in the gray areas are based on multiple samples (cf. Figure 6.7). Systematic errors of the measurement method are discussed in 3.2.1.

| $\dot{m}_{total}$ kg/h | $\phi_m$ - | $x_{k,3}$ nm | $x_{50,3}$ nm | $x_{10,3}$ nm | $x_{90,3}$ nm | $x_{g,3}$ nm | $\sigma_{g,3}$ |
|---|---|---|---|---|---|---|---|
| 3.04 | 0.05 | $53.9 \pm 1.0$ | $46.4 \pm 0.9$ | $21.0 \pm 0.5$ | $80.5 \pm 1.3$ | $43.5 \pm 0.75$ | $1.73 \pm 0.01$ |
| 3.04 | 0.1 | $63.4 \pm 0.1$ | $46.9 \pm 0.1$ | $18.7 \pm 0.3$ | $83.0 \pm 1.2$ | $42.4 \pm 0.5$ | $1.79 \pm 0.02$ |
| 3.04 | 0.15 | $60.5 \pm 1.5$ | $41.8 \pm 0.5$ | $17.0 \pm 0.4$ | $79.0 \pm 0.7$ | $38.3 \pm 0.4$ | $1.82 \pm 0.02$ |
| 3.04 | 0.2 | $23.0 \pm 0.1$ | $26.9 \pm 0.2$ | $15.7 \pm 0.1$ | $74.0 \pm 0.5$ | $30.6 \pm 0.3$ | $1.80 \pm 0.01$ |
| 4.55 | 0.1 | $60.1 \pm 1.3$ | $38.0 \pm 0.7$ | $17.4 \pm 0.3$ | $78.3 \pm 0.5$ | $37.3 \pm 0.5$ | $1.80 \pm 0.01$ |
| 6.07 | 0.05 | $70.4 \pm 0.4$ | $40.3 \pm 0.5$ | $17.2 \pm 0.15$ | $87.2 \pm 0.1$ | $39.3 \pm 0.2$ | $1.90 \pm 0.01$ |
| 6.07 | 0.1 | $23.1 \pm 0.1$ | $28.5 \pm 0.1$ | $15.9 \pm 0.6$ | $76.5 \pm 0.1$ | $32.2 \pm 0.1$ | $1.82 \pm 0.01$ |
| 6.07 | 0.15 | $23.3 \pm 0.1$ | $24.7 \pm 0.1$ | $15.2 \pm 0.1$ | $53.7 \pm 0.7$ | $26.6 \pm 0.1$ | $1.66 \pm 0.01$ |
| 6.07 | 0.2 | $20.6 \pm 0.9$ | $21.6 \pm 0.7$ | $13.7 \pm 0.4$ | $41.1 \pm 2.9$ | $23.1 \pm 0.8$ | $1.62 \pm 0.04$ |
| 6.07 | 0.3 | $21.9 \pm 0.7$ | $22.7 \pm 0.2$ | $14.6 \pm 0.3$ | $43.3 \pm 2.2$ | $24.4 \pm 0.7$ | $1.63 \pm 0.01$ |
| 9.11 | 0.2 | $19.8 \pm 1.1$ | $20.1 \pm 1.0$ | $13.2 \pm 0.6$ | $32.7 \pm 1.7$ | $21.0 \pm 1.0$ | $1.54 \pm 0.03$ |

Due to decreasing fractions of the bigger particles, $x_{g,3}$ declines upon increasing $\phi_m$ and $\dot{m}_{total}$. For $\dot{m}_{total} = 3.04$ kg/h the strongest variation is observed between $\phi_m$ of 0.15 and 0.2. In comparison, for a $\dot{m}_{total}$ of 6.07 kg/h the influence of $\phi_m$ continuously decreases from 0.05 to 0.2. Increasing $\phi_m$ beyond this value, did not induce any further significant changes in $x_{g,3}$.

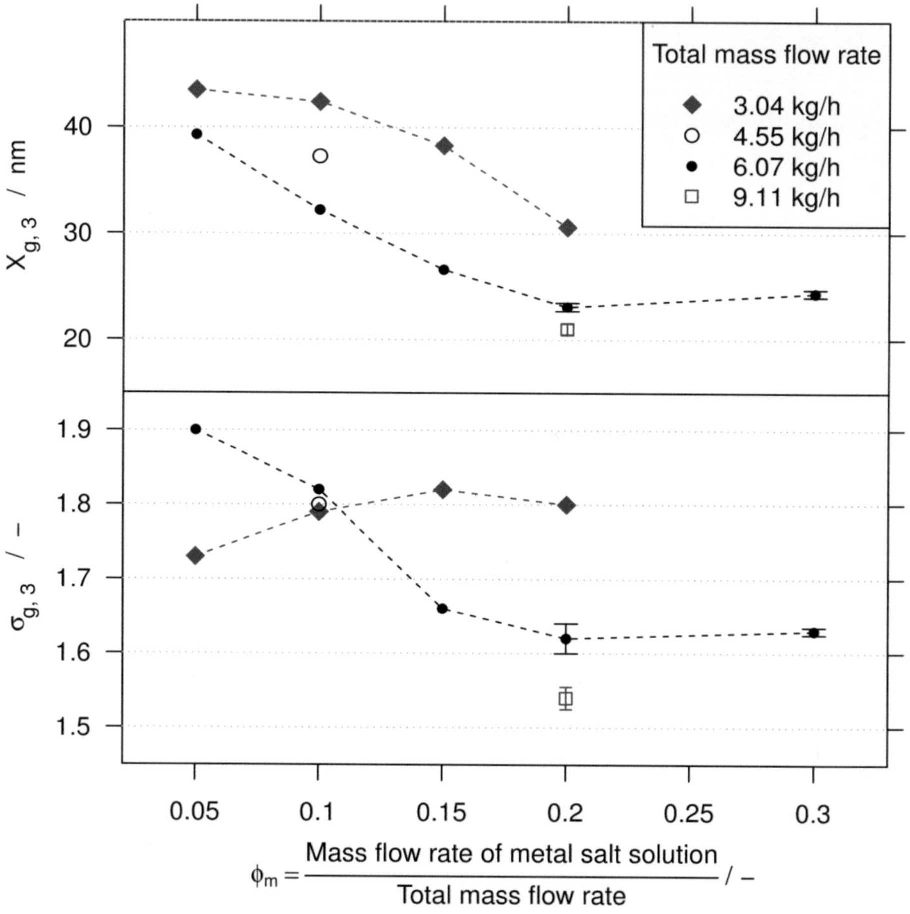


**Figure 6.7:** Geometric mean ($x_{g,3}$) (top) and geometric standard deviation ($\sigma_{g,3}$) (bottom) as function of the mixing value ($\phi_m$) and the total mass flow rate ($\dot{m}_{total}$). The connecting lines between the data points serve solely for clarity purposes. The error bars correspond to one standard deviation based on different samples which were either taken from five independent experimental runs and/or taken after different times from a long-term experimental run.

Using the geometric standard deviation ($\sigma_{g,3}$) as a measure of the width of the particle size distributions, a broadening of the particle size distribution is observed upon increasing $\phi_m$ for the lowest total mass flow rate (3.04 kg/h) investigated, with a faint maximum at $\phi_m = 0.15$. At the higher mass flow rate in this investigation ($\dot{m}_{total} = 6.07$ kg/h), the geometric standard deviation continuously decreases up to $\phi_m = 0.2$ with no further significant changes regarding the mean (of the standard deviation) beyond this value of $\phi_m$. However, the reproducibility of the particle size distribution—as given by the standard deviation of the geometric standard deviation—was much better for $\phi_m = 0.3$ than for $\phi_m = 0.2$. This is discussed in detail below.

The observed progression of the geometric standard deviation ($\sigma_{g,3}$) is attributed to changes in the ratio between the smaller and larger particles. At low total mass flow rates and mixing values (the latter meaning low mass flow rates of the salt stream, $\dot{m}_{salt}$), the fraction of the larger particles is dominant with the result of fairly narrow particle size distributions. Increasing $\phi_m$ and/or $\dot{m}_{total}$ caused a broadening of the particle size distributions through an increase in the smaller particle fraction. Further increases in $\phi_m$ and/or $\dot{m}_{total}$ lead to the domination of the smaller particles with the result of again narrower distributions.

Although the progression described above could be shifted to smaller or larger values due to the uncertainty of the material properties, the general progression is not effected. Due to this uncertainty, a complex refractive index was chosen for which it is reasonable to assume that it is more likely that the fraction of the larger particles is overestimated than underestimated (for details see Section 3.2.1). Taking this into account, with reasonable high $\phi_m$ ($\geq 0.2$) and $\dot{m}_{total}$ ($\geq 6.07$), the formation of the fraction of the larger particles could be almost fully suppressed, presumably due to the higher values of $\dot{m}_{salt}$ which correlates to a reduction of the heating period of this stream before the mixing with supercritical water stream. Furthermore, increasing $\dot{m}_{total}$ and changing $\phi_m$ can surely enhance effective and homogenous mixing with the result of more uniform particle formation. The further decrease of the values $x_{g,3}$ and $\sigma_{g,3}$ upon the increase of $\dot{m}_{total}$ to 9.11 kg/h at a constant $\phi_m$ of 0.2 could be also attributed to the decreased residence time. A similar trend—but for primary particles measured with TEM—was reported by Sue et al. (2011) for iron oxide nanoparticles also synthesized from iron(III) nitrate under same temperature and pressure conditions but slightly lower product concentration than used here (0.0033 mol/L prior to quenching vs. 0.0055 mol/L). In their study, the residence times varied from 2 ms to 2 s with corresponding mean primary particle sizes from 4 to 7 nm.

In this context, it is important to note that the formation of bimodal distributions was not observed in the hydrothermal synthesis of cerium oxide nanoparticle dispersions under nearly identical conditions (cf. 5). Therefore, material specific properties surely play a pivotal role. It is reasonable to assume that the reaction mechanisms of the hydroxide and oxide formation are much more complicated than the following overall reactions suggest (Adschiri and Arai, 2002):

$$\mathrm{Me^{z+} + zOH^- \rightleftarrows Me(OH)_z} \tag{6.1}$$

$$\mathrm{Me(OH)_z \rightleftarrows MeO_{z/2} + \frac{z}{2}H_2O} \tag{6.2}$$

where Me stands for metal. Furthermore, the ratios between the different time scales of the involved processes are dependent on the material specific properties. A material system which is especially sensitive to variations in the reaction and solvent medium might be of great interest for the ex situ assessment of the mixing conditions. The results presented for the iron oxide nanoparticle dispersions as well as the knowledge of the complexity of possible iron compounds in aqueous solutions (Cornell and Schwertmann, 2003) imply that the investigations were undertaken with a highly sensitive material system.

In light of this, it becomes necessary to address the question of the degree of control of the process conditions necessary to obtain dispersions with reasonably steady product properties out of the continuous process. This question was addressed by carrying out several repetitive runs under conditions where the formation of the fraction of the larger particles was strongly reduced ($\phi_m \geq 0.2$ and $\dot{m}_{total} \geq 6.07$ kg/h).

Representative results are shown in Figure 6.8. Here, characteristic values of the cumulative particle size distribution ($Q_3$) are displayed as function of the pressure difference ($\Delta p$) between the pressure indicators after the reactor and before the mixing unit. Shortly after the start of the continuous particle synthesis, particles deposit at the entry of the metal salt stream into the mixing unit accompanied by the reduction of the cross-sectional area resulting in an increase of the mean flow velocity. Due to this fact, a pressure difference across the mixing unit builts up. Examination of the characteristic values of the particle size distribution in relation to the pressure difference revealed a strong dependence especially between the $x_{90,3}$ and the pressure difference. However, the changes in the particle size distributions decreased upon the increase of the pressure difference. Beyond a pressure difference of 0.25 MPa only marginal changes were measurable anymore. It is important to note that pressure differences below 0.25 MPa could only be realized with a newly implemented mixing unit. Furthermore, the influence of the particle deposits at the entry of the metal salt stream into the mixing unit—ascertainable by measuring the pressure difference across the mixing unit—on the particle size distribution was much stronger during the run-in period than fluctuations in pressure (mean variation from the set value up to 0.5 MPa) and temperature (mean variation from the set value up to 2 K).

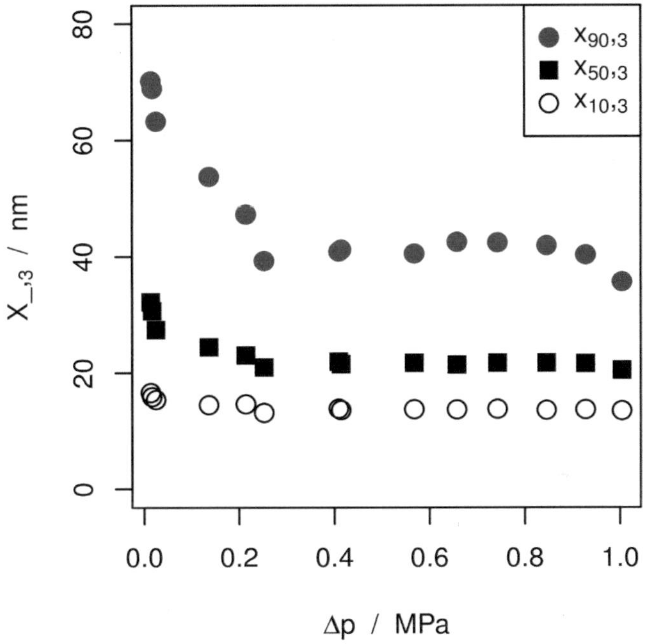


**Figure 6.8:** Characteristic values of the cumulative particle size distributions ($Q_3$) of dispersions synthesized at $\phi_m = 0.2$ and $\dot{m}_{total} = 6.07$ kg/h as a function of the pressure differences ($\Delta p$) between the pressure indicators after the reactor and before the mixing unit.

A slow increase of the pressure difference was always recorded in experiments that were carried out under these conditions over the course of several hours. However, the particle deposits causing the pressure difference could be strongly reduced by applying pressure surges after the run and the switch to pure water. Therefore, the subsequent experiment was started again at a low pressure difference. Experiments carried out at an even higher $\phi_m$ proved to be more stable as already reported and discussed above. However, a disadvantage of higher $\phi_m$ is the heat supply at higher temperature levels in order to achieve the same mixing temperature. Preferably, this should be avoided due to energetic reasons as well as due to increased demands on the construction material.

To further characterize the particles obtained under different flow conditions, selected samples were examined by means of transmission electron microscopy to investigate the size distribution of the primary particles. In Figure 6.9 representative TEM images of (i) a sample with a strong bimodal distribution of the dispersed particles (top) and (ii) of a specimen in which the fraction of bigger particles was essentially not present (bottom) are shown. Particle size distributions corresponding to these flow conditions are shown in Figure 6.6. Inspection of the TEM images of the first sample revealed also the presence of two very different fractions of particles which appeared quite seggregated. One fraction of particles of fairly spherical shapes had generally diameters in the range of 4 to 10 nm (average around 7 nm) and the other fraction of irregularly shaped particles in the range of 15 to 50 nm (average around 30 nm). On the other hand, overview images of the latter sample (e.g. bottom left) revealed a homogeneous particle distribution. HRTEM images revealed the presence of fairly spherical particles in the range of 4 to 10 nm similar to the fraction of the smaller particles observed for the first sample. Similar results were reported by Sue et al. (2010) using a micro T-mixer.

The following overall picture emerges combining the results of the structural investigations, the color of the dispersions, the results using the disc centrifuge and the inspection of the TEM images: electrostatically stabilized iron oxide nanoparticle dispersions can be synthesized in a continuous process. The synthesized dispersions show bimodal size distributions of the particles/agglomerates, with one mode around 20 nm and one around 60 nm. The relation of these two modes in the particle size distribution can be changed through varying the flow conditions, up to the point where the second mode (larger particles) can be almost totally suppressed. At the level of primary particles, also two fractions of particles are formed: The fraction of the larger particles consisting of hematite nanoparticles with dimensions around 30 nm (size range from 15 to 50 nm). The other fraction consists mostly of particle agglomerates and/or aggregates formed by primary particles generally with an average size of around 7 nm (size range 4 to 10 nm). At least a part of the particles is of a different structure than hematite, presumably ferrihydrite and/or maghemite. Analogous to what was observed for the size distribution of particles in dispersion,

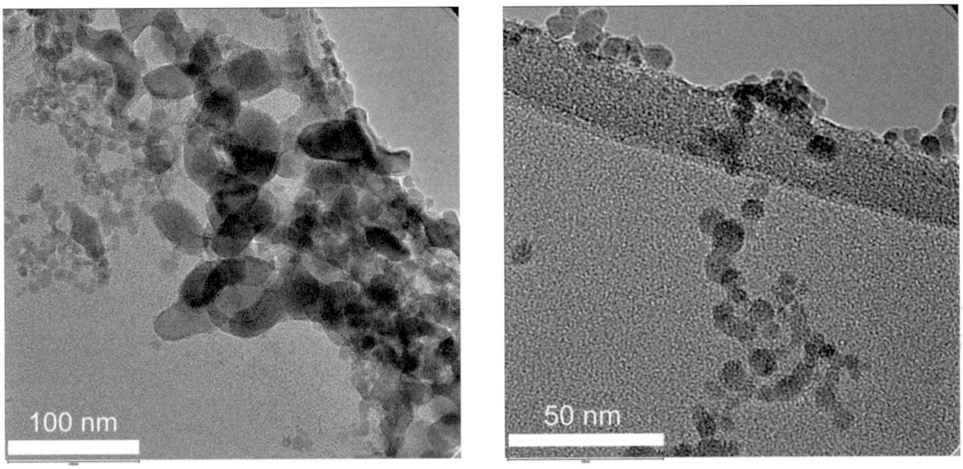


**(a)** $\phi_m = 0.2$, $\dot{m}_{total} = 3.04$ kg/h

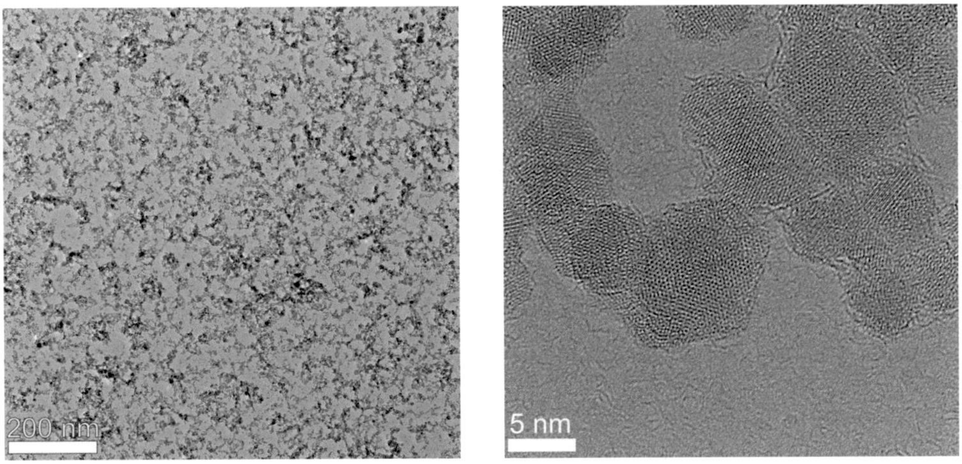


**(b)** $\phi_m = 0.2$, $\dot{m}_{total} = 9.11$ kg/h

**Figure 6.9:** TEM images of particles synthesized from aqueous iron nitrate solutions with $c_{\text{product Fe}} = 0.0055$ mol/L under different flow conditions.

the fraction of the larger primary particles can be suppressed down to a point where essentially neither large primary particles nor particle agglomerates $\geq$ 50 nm are present by increasing $\phi_m$ and/or $\dot{m}_{total}$.

Furthermore, the results of the investigations on varying the flow conditions provide useful insights for the optimization of the mixing process. In addition, by using the stable iron nitrate solutions as a starting material in combination with the powerful and easy to use tool of the disc centrifuge, a feasible method to characterize and assess the mixing ex situ is given.

## 6.2 Dispersions synthesized from different starting materials

Another important way to influence the product properties is the variation of the chemical composition of the metal salt stream. This avenue was explored by (i) changing the ratio between iron(III) nitrate and iron(II) acetate and (ii) by the addition of acetic acid and/or nitric acid. First, the results from experiments carried out using iron(II) acetate as the only iron salt are presented. Second, the most important results of the systematic investigations on the influence of the ratio between the two iron salts are shown and discussed. In these experiments, the total amount of acetate and nitrate was also varied by addition of the respective acids.

### 6.2.1 Dispersions synthesized from aqueous iron acetate solutions

Particle suspensions synthesized from iron acetate solutions ($pH_{salt} = 4$) appeared with a darkish color. Furthermore, a fraction of the particles sedimented slowly but noticeably. However, after the sedimentation of these particles, the supernatant was still slightly darkish (see Figure 6.10), suggesting that the suspensions consisted of two fractions of particles with very different dimensions. This was confirmed by analyzing the particles via TEM. These investigations revealed the presence of two particle fractions, one containing particles with diameters in the range of 50 - 200 nm and the other one containing particles with diameters below 10 nm (see Figure 6.11). Furthermore, the TEM images suggest that larger particles mostly formed aggregates and/or agglomerates whereas the smaller particles existed predominantly as distinct particles. The finding of two fractions of particles with very different dimensions when using iron(II) acetate as starting material is in agreement with the work by Cabanas and Poliakoff (2001).

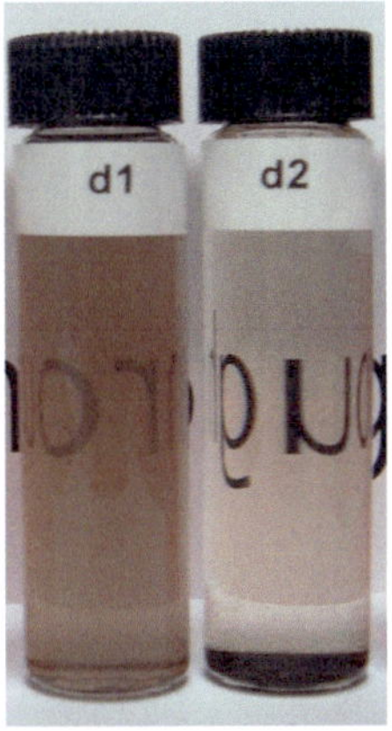


**Figure 6.10:** Suspensions synthesized from aqueous iron acetate solutions before (left) and after allowing sufficient time for sedimentation (right). No further changes were observable over a period of several months. Synthesis conditions: $\phi_m = 0.1$, $\dot{m}_{total} = 6.07$ kg/h, $c_{\text{product Fe}} = 0.001$ mol/L.

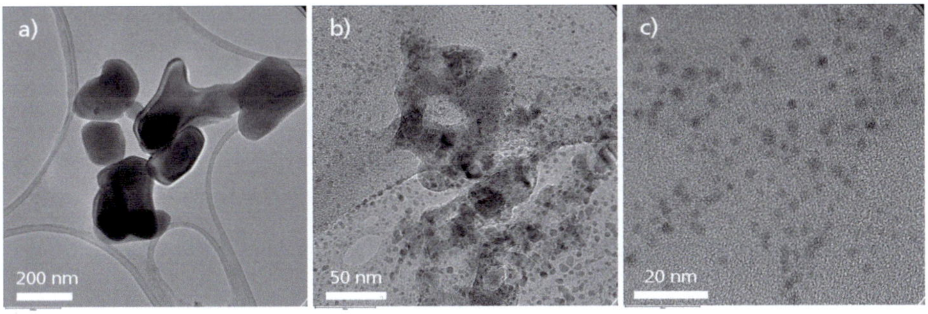


**Figure 6.11:** TEM images of the particles from the suspension shown in Figure 6.10 at different magnifications.

The crystalline structures of the dried particles from the sample shown in Figure 6.10 were identified as magnetite via PXRD analysis. A mean crystallite size of 48 nm was calculated using a whole pattern fit of the diffraction pattern (see Figure E.21 on page 256). However, the presence of a bimodal particle size distribution opened the possibility of obtaining stable iron oxide nanoparticle suspensions with discrete single particles of sizes below 10 nm by suppressing the fraction of the larger particles.

### 6.2.2 Dispersions synthesized from acidified aqueous iron acetate solutions using nitric acid

While exploring the possibility of influencing the particle size distribution by changing the composition of the iron salt solution, at first solutions with a low iron content ($c_{\text{product Fe}}$ = 0.001 mol/L) were used to reduce the amount of material used and to reduce the risk of clogging due to the deposition of particles. In experimental runs that were carried out using iron(II) acetate from the same packaging unit as used for the suspension shown in Figure 6.10, highly transparent and stable dispersions could be obtained for certain amounts of added nitric acid. However, the color of these transparent dispersions was yellow-orange. After concentrating the dispersion to an iron content similar to the ones obtained using iron nitrate solutions, the color of the dispersion was similar to those (see Figure 6.2). These highly transparent dispersions were obtained when 0.002 mol/L of nitric acid was added. The addition of larger amounts of nitric acid (0.02 mol/L) was accompanied by an increase in the turbidity indicating the presence of larger particles.

A more in depth examination of the as-synthesized dispersions by means of the differential sedimentation analysis revealed that the particle size distributions of the highly transparent dispersions ($c_{HNO_3}$ = 0.002 mol/L) were very narrow with less than 5% (by weight) of the dispersed particles larger than 17 nm for $\dot{m}_{total}$ = 6.07 kg/h or larger than 22 nm for $\dot{m}_{total}$ = 3.04 kg/h, with modes slightly larger than 13 nm (calculated assuming the density of ferrihydrite) (see Figure 6.12 and Table 6.5). Significant differences between the particle size distributions obtained at two different flow rates were only observed for particles in the fraction of the largest 20%. For dispersions synthesized with $c_{HNO_3}$ = 0.02 mol/L, relatively narrow main peaks with a noticeable tail in the direction of larger particles and a second peak with a mode around 160 nm containing about 10% of the particles were revealed by the analysis using the disc centrifuge. Differences between the two different set total mass flow rates were mainly related to the position of the main peak, which was shifted to smaller particle sizes (by around 1.5 nm) upon increasing $\dot{m}_{total}$ from 3.04 kg/h to 6.07 kg/h.

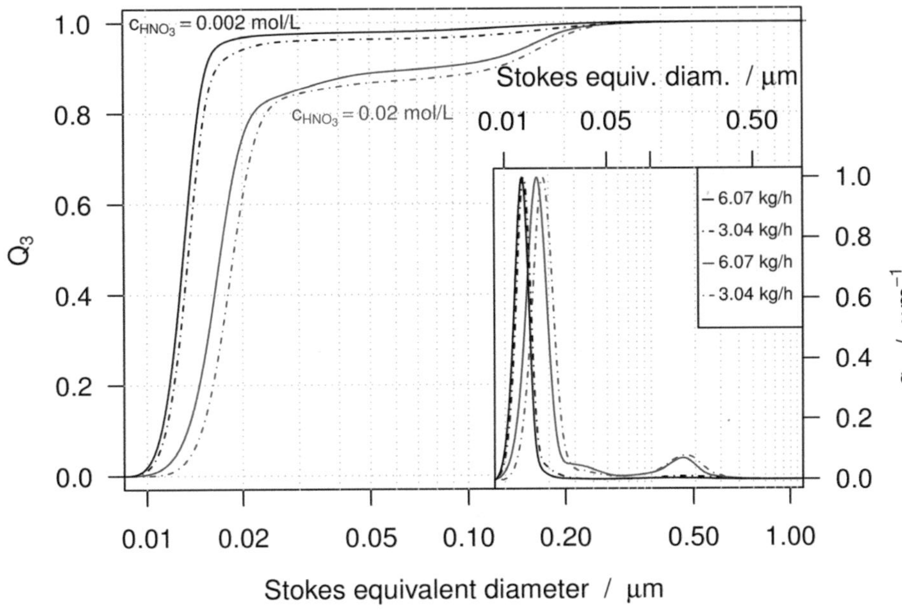


**Figure 6.12:** Particle size distributions ($Q_3$ and $q_3$) of dispersions synthesized from iron(II) acetate at two different total mass flow rates ($\dot{m}_{total}$) and upon the addition of different amounts of nitric acid with $\phi_m$ = 0.2 and $c_{\text{product Fe}}$ = 0.001 mol/L).

Comparing the particle size distributions obtained with the acidified iron(II) acetate solutions (Figure 6.12) with the ones from iron(III) nitrate solutions (Figure 6.6), it is immediately striking that the peak of the smaller particles is generally narrower for the former. However, to exclude the possibility that this was only due to the higher iron content (5.5 times higher for the iron(III) nitrate solutions), an experimental run was carried out with exactly the iron content that was used for the previous study with the same composition of the starting material and the flow conditions that produced the narrowest and smallest particle size distributions. A particle size distribution of a product suspension obtained with the larger iron content ($c_{\text{product Fe}}$ = 0.0055 mol/L) is shown in Figure 6.13 in combination with particle size distributions of two product suspensions with the lower iron content ($c_{\text{product Fe}}$ = 0.001 mol/L) obtained at two independent experimental runs. Comparing the two latter ones, a very high reproducibility is observed with only marginal differences for the very small fraction of the particles larger than 17 nm. In the case of the larger iron content, the shapes of the particle size distributions were preserved with the only difference that they were shifted to slightly larger particle sizes (difference of the modes 1.3 nm).

**Table 6.5:** Characteristic values of the particle size distributions synthesized from acidified iron(II) acetate at two different total mass flow rates ($\dot{m}_{total}$) and upon the addition of different amounts of nitric acid with $\phi_m = 0.2$. The corresponding particle size distributions are shown in Figure 6.12 and Figure 6.13. Due to a high uncertainty in material properties of the particles, the systematic error for particle size distributions with a substantial fraction of larger particles could be large; affected values are enclosed in parentheses. Systematic errors of the measurement method are discussed in 3.2.1.

| $\dot{m}_{total}$ kg/h | $c_{\text{product Fe}}$ mol/L | $HNO_3$ mmol/L | $pH_{salt}$ - | $pH_{product}$ - | $x_{k,3}$ nm | $x_{50,3}$ nm | $x_{10,3}$ nm |
|---|---|---|---|---|---|---|---|
| 3.04 | 0.001 | 2 | 3.0 | 3.7 | $13.4 \pm 0.6$ | $13.6 \pm 0.6$ | $11.5 \pm 0.4$ |
| 3.04 | 0.001 | 20 | 2.0 | 2.4 | $18.2 \pm 0.3$ | $18.8 \pm 0.3$ | $14.7 \pm 0.3$ |
| 6.07 | 0.001 | 2 | 3.0 | 3.5 | $13.1 \pm 0.2$ | $13.0 \pm 0.2$ | $11.1 \pm 0.1$ |
| 6.07 | 0.001 | 20 | 2.0 | 2.4 | $16.7 \pm 0.2$ | $17.1 \pm 0.2$ | $13.4 \pm 0.1$ |
| 6.07 | 0.001 | 2 | 3.0 | 3.5 | $13.3 \pm 0.3$ | $13.1 \pm 0.3$ | $11.0 \pm 0.2$ |
| 6.07 | 0.0055 | 5 | 3.0 | 3.2 | $14.6 \pm 0.3$ | $14.5 \pm 0.3$ | $12.3 \pm 0.2$ |

| $\dot{m}_{total}$ kg/h | $c_{\text{product Fe}}$ mol/L | $HNO_3$ mmol/L | $pH_{salt}$ - | $pH_{product}$ - | $x_{90,3}$ nm | $x_{g,3}$ nm | $\sigma_{g,3}$ |
|---|---|---|---|---|---|---|---|
| 3.04 | 0.001 | 2 | 3.0 | 3.7 | $16.7 \pm 1.2$ | $15.0 \pm 0.8$ | $1.65 \pm 0.1$ |
| 3.04 | 0.001 | 20 | 2.0 | 2.4 | $(118.3 \pm 5.9)$ | $(24.2 \pm 0.8)$ | $(2.11 \pm 0.04)$ |
| 6.07 | 0.001 | 2 | 3.0 | 3.5 | $15.3 \pm 0.4$ | $13.7 \pm 0.3$ | $1.44 \pm 0.04$ |
| 6.07 | 0.001 | 20 | 2.0 | 2.4 | $(71.6 \pm 7.0)$ | $(21.6 \pm 0.5)$ | $(2.05 \pm 0.1)$ |
| 6.07 | 0.001 | 2 | 3.0 | 3.5 | $15.5 \pm 0.6$ | $13.5 \pm 0.4$ | $1.36 \pm 0.06$ |
| 6.07 | 0.0055 | 5 | 3.0 | 3.2 | $17.3 \pm 0.6$ | $15.2 \pm 0.4$ | $1.43 \pm 0.03$ |

The difference observed in the width of the particle size distributions ($\sigma_{g,3} = 1.55$ for product dispersions obtained from iron(III) nitrate solutions under the most favorable flow conditions investigated and $c_{\text{product Fe}} = 0.0055$ mol/L, and $\sigma_{g,3} = 1.43$ for the same conditions and iron content but using acidified iron(II) acetate solutions) points to differences in the particle formation process. This was further explored by comparing the distribution of the primary particles investigated by means of TEM.

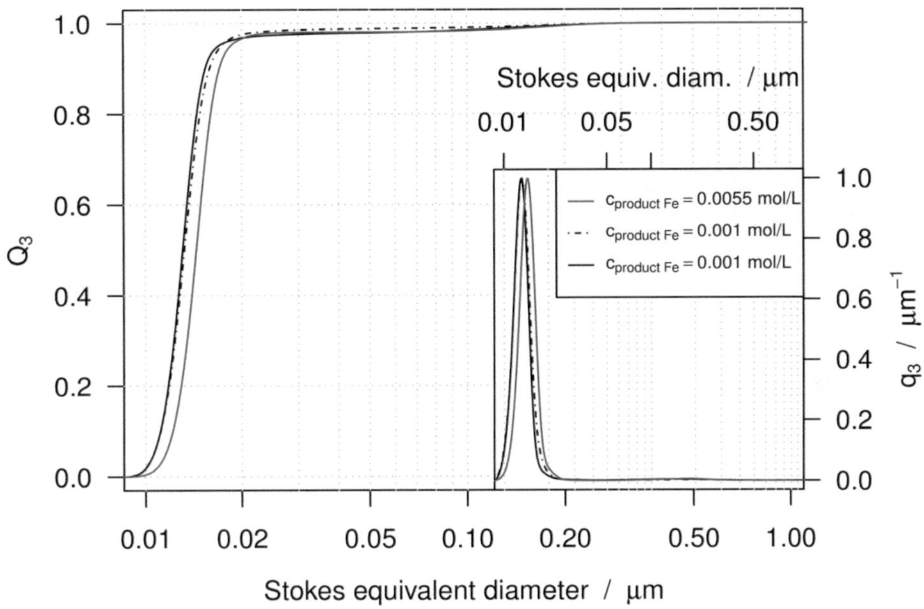


**Figure 6.13:** Particle size distributions ($Q_3$ and $q_3$) of dispersions synthesized from acidified iron(II) acetate solutions at $\dot{m}_{total} = 6.07$ kg/h and $\phi_m = 0.2$. For the dispersions with $c_{\text{product Fe}} = 0.001$ mol/L, 2 mmol/L nitric acid was added to the starting solutions; for the dispersion with $c_{\text{product Fe}} = 0.0055$ mol/L, 5 mmol/L nitric acid was added to the starting solution.

Representative images of particles from the highly transparent dispersions obtained with acidified iron(II) acetate solutions applying two different $\dot{m}_{total}$ are shown in Figure 6.14.

In the case of the lower mass flow rate (3.04 kg/h), two distinctive particle fractions were observable. The dominant fraction consisted of fairly spherical particles with a narrow size distribution of diameters generally between 3.5 and 6.5 nm (mean value about 5 nm). The other fraction consisted of highly spherical particles in the size range of 20 to 80 nm (mean value about 40 nm). The spherical shape in comparison to the irregular shapes of the larger particles observed for particles synthesized from iron nitrate solutions suggests that the latter were formed predominantly through coalescence of smaller particles whereas the former through diffusion and incorpo-

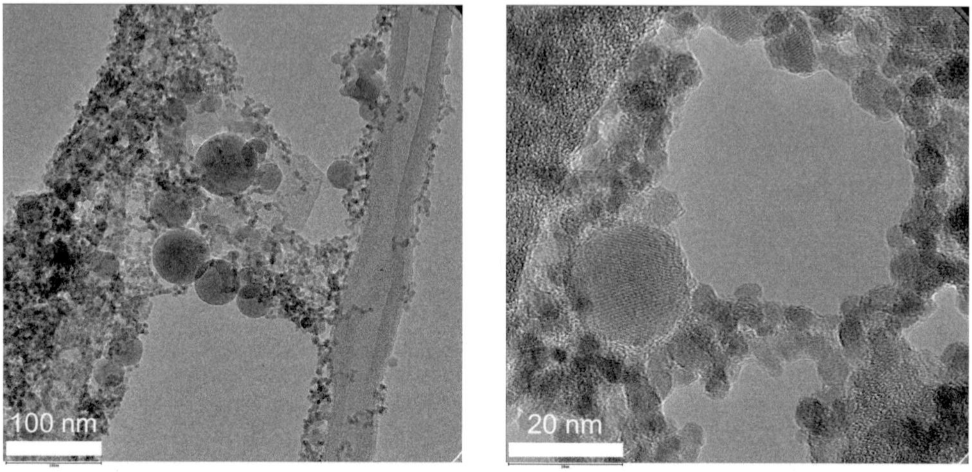


**(a)** $\phi_m = 0.2$, $\dot{m}_{total} = 3.04$ kg/h

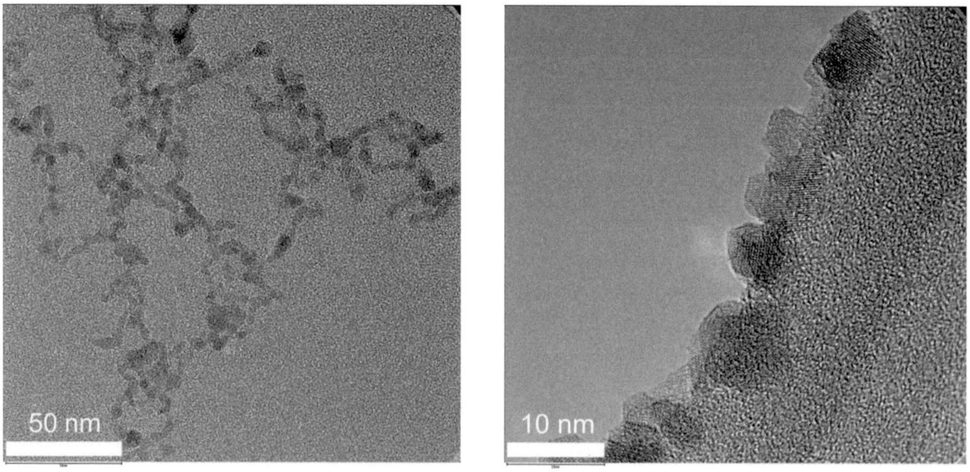


**(b)** $\phi_m = 0.2$, $\dot{m}_{total} = 6.07$ kg/h

**Figure 6.14:** TEM images of particles synthesized from aqueous iron(II) acetate solutions with 0.002 mol/L nitric acid with $c_{\text{product Fe}} = 0.001$ mol/L.

ration of iron species at the surface. In the case of particles synthesized at $\dot{m}_{total}$ = 6.07 kg/h, for which only a marginal portion of particles above 20 nm was observed for the mass distribution of the dispersion, only primary particles very similar to the ones of the dominant fraction for the sample with a $\dot{m}_{total}$ = 3.04 kg/h were present (size range from 3.5 nm to 6.5 nm, mean value about 5 nm).

The fraction of smaller primary particles synthesized from acidified iron(II) acetate showed a narrower size range and the particles were on average smaller than the corresponding fraction of particles synthesized from iron(III) nitrate. The difference for the two material systems observed in both particle fractions (the small fraction described here and the fraction of larger particles described above) points to an inhibition of the coalescence of smaller particles. This is maybe due to a better stabilization of the particles caused by a different environment and/or surface chemistry. Although this could not be investigated in situ, investigations on the structure and the stability ex situ could provide some insights.

Raman spectroscopic investigations were undertaken with a concentrated iron oxide nanoparticle dispersion synthesized with $\phi_m$ = 0.2 and $\dot{m}_{total}$ = 6.07 kg/h using iron(II) acetate with the addition of 0.002 mol/L $HNO_3$ and with $c_{\mathrm{product\,Fe}}$ = 0.001 mol/L (cf. left curves in Figure 6.12). In Figure 6.15, the Raman spectra of the dispersions measured at two different laser power settings are shown. Neglecting the line at 1048 $cm^{-1}$, which is attributed to the total symmetric stretching vibration of the nitrate ion, the spectra are dominated by a broad and asymmetric band with a maximum at 718 $cm^{-1}$. Further, two broad, weak features can be observed at 365 $cm^{-1}$ and 497 $cm^{-1}$, respectively, and one very broad peak which might be a superposition of two bands from 1250 to 1520 $cm^{-1}$. Another weak band is observable at 914 $cm^{-1}$. However, no structural transformations were recognized upon using a higher laser power in order to improve the signal-to-noise ratio of the spectra. The shape and position of the dominant features as well as the very broad band in the wavenumber range greater than 1200 $cm^{-1}$ points to the structure of ferrihydrite. A reference spectrum of ferrihydrite which is in accordance with literature data (Mazzetti and Thistlethwaite, 2002) and the respective peak positions are also displayed at the top of Figure 6.15. Especially, the very asymmetric shape of the strongest band is in good agreement with the observed spectra of hydrothermally synthesized particles. However, the unidentified band at 914 $cm^{-1}$ already seen in spectra of Figure 6.15 also appears in spectra of samples measured as dispersion (Figure 6.18). This mode is absent in case of dried sample(s) as discussed previously. The possibility of structural transformations upon (freeze-)drying was stated by Peak and Regier (2012).

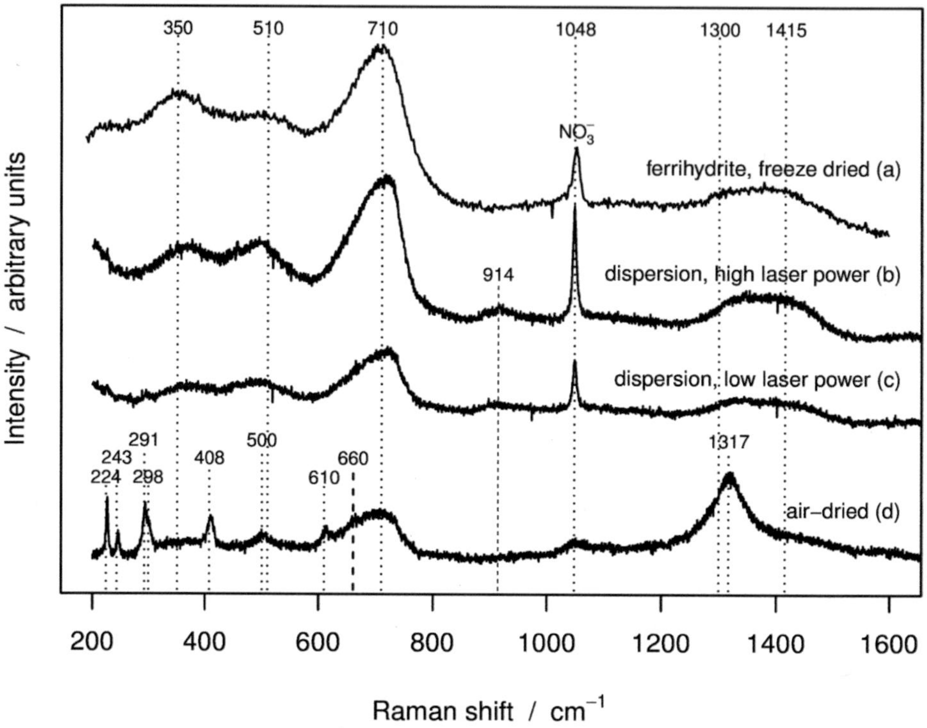


**Figure 6.15:** Raman spectra from: a) reference sample 2-line ferrihydrite prepared according to Schwertmann and Cornell (2000); b) iron oxide nanoparticle dispersion synthesized with $\phi_m = 0.2$ and $\dot{m}_{total} = 6.07$ kg/h using iron(II) acetate with the addition of 0.002 mol/L $HNO_3$ and with $c_{\text{product Fe}} = 0.001$ mol/L, measured as a concentrated dispersion ("hanging drop") with a laser power of 2.0 mW; c) same sample as b) but measured with a laser power of 0.5 mW; d) as b) measured after air-drying. The spectra are shifted vertically for clarity.

Upon drying under ambient conditions, the characteristic modes of hematite appeared and the band at 914 $cm^{-1}$ disappeared (see spectrum at the bottom of Figure 6.15). Due to the overlapping of different bands in the region around 690 $cm^{-1}$, it is not possible to resolve if the broadening of the band at 710 $cm^{-1}$ in direction of lower wavenumbers, indicating the presence of maghemite, is (i) only due to maghemite-like shell structures of nanoparticulate hematite or (ii) due to formation of maghemite nanoparticles upon drying. Furthermore, it is not possible to infer from this convoluted spectrum whether the ferrihydrite nanoparticles were completely or only partially transformed into hematite and maghemite upon drying.

To further explore the differences between the suspended particles obtained from the acidified iron(II) acetate solutions in comparison to the ones obtained from iron(III) nitrate solutions, their stability in the aqueous media was investigated by means of electrophoretic lightscattering. In the case of using iron(III) nitrate as starting material, electrophoretic mobilities of $(3.75 \pm 0.14) \times 10^{-8}$ m$^2$/Vs and $\zeta$ potentials of $(47.8 \pm 1.8)$ mV (derived by applying the model of Smoluchowski) were determined irrespective of the flow conditons. Significantly higher values (electrophoretic mobilities of $(4.19 \pm 0.04) \times 10^{-8}$ m$^2$/Vs and a $\zeta$ potential of $(53.4 \pm 0.5)$ mV were obtained for the sample which was synthesized from acidified iron(II) acetate solutions with the same iron content in the product stream ($c_{\text{product Fe}}$ = 0.0055 mol/l). Similar values were deduced for the dispersions with a lower iron content ($c_{\text{product Fe}}$ = 0.001 mol/l). This might be an indication of differences in the surface chemistry of the particles which could influence the stability of the particles, also under supercritical water conditions, with the result of differences in the particle formation process.

Following the finding of highly stable dispersions with very narrow particle size distributions accompanied by low fluctuations in the process conditions leading to a high reproducibility of the resulting dispersions, a further increase of the iron content was explored. However, when carrying out the experimental run with increased iron content of the product suspension, a new packaging unit of the iron(II) acetate had to be used. It was immediately striking that the color of the freshly opened unit was different from the color of iron(II) acetate from the first unit (beige/gold compared to a brownish red of the first unit used)*. Furthermore, aqueous solutions of the newly opened iron(II) acetate with the same concentration and added amount of nitric acid exhibited a higher pH value. Experimental runs using these solutions did not give comparable results to the previous runs and proved to be unfavorable with respect to process stability and the clogging of the tubings. The strong red color (it is known that the color of iron(II) acetate varies between beige and slightly red) of

*It is worth noting that the brownish red color was already observed at the time this unit was opened, directly after breaking the air tight packaging cover; there were no apparent color changes afterwards. To minimize the exposure to air the iron(II) acetate was kept in a glove box with an argon atmosphere, and before each experiment, a sufficent amount was bottled air tight in the glove box and was opened under normal atmosphere only once to prepare the aqueous solution.

the first packaging unit and the different buffering behavior suggest that the iron(II) was already partially oxidized to iron(III) and an appreciable amount of the expected acetate was not present, either due to the evaporation of acetic acid and/or contamination. Ion chromatography analysis of an old solution of iron(II) acetate initially used revealed the presence of further two small peaks beside the dominant acetate peak, one of which could be attributed to formiate. The occurence of these two small but clearly observable peaks was not detected for any other packing unit investigated (see Figure E.16 on page 250).

### 6.2.3 Dispersions synthesized from mixtures of iron(III) nitrate and iron(II) acetate

The promising results with the initial iron acetate solutions and the possibility of partially oxidized iron(II) and a deficiency in the acetate content prompted an investigation into the influence of using mixtures of iron(II) acetate and iron(III) nitrate on the product dispersions. The total amounts of acetate and nitrate were varied by addition of the respective acids.

For the purpose of finding compositions of the starting solution that lead to dispersions with similar narrow size distributions as observed above (cf. Section 6.2.2), experiments were carried out using various compositions at a low iron content ($c_{\text{product Fe}}$ = 0.001 mol/L). The ratio between iron(II) and iron(III) was varied by addition of iron(II)acetate and iron(III)nitrate, respectively, from 45:55 up to 75:25. Furthermore, the total amount of acetate was varied by the addition of acetic acid. The total acetate content is given as "acetate factor" with respect to the iron content (mol/mol). The acetate factor was varied from 1.7 to 2.1. Additionally, different amounts of nitric acid were added (from 0 to 4 mmol/L). The compositions of the starting solutions investigated are summarized in Table 6.6 and selected results of these screening experiments are presented in Figure 6.16.

Inspection of Figure 6.16 reveals that it is possible to obtain narrow particle size distributions by different parameter combinations regarding the composition of the starting solution. For easier comparison, the narrowest distribution from Figure 6.13 is shown as a reference in Figure 6.16. The differences between the particle size distributions displayed in Figure 6.16 are small regarding both the mode (range for all

**Table 6.6:** Overview of the compositions of the starting solutions used to synthesize dispersions from mixtures of iron(II) acetate and iron(III) nitrate.

| | Addition of $HNO_3$ | | | |
|---|---|---|---|---|
| Fe(II):Fe(III) | 0 mmol/L | 2 mmol/L | 4 mmol/L | |
| 45:55 | 2.1 | 2.1 | 2.1 | |
| 55:45 | 1.7/1.9/2.1 | 1.7/1.9/2.1 | 1.7/1.9/2.1 | Acetate |
| 65:35 | 1.7/1.9 | 1.7/1.9 | 1.7/1.9 | factor |
| 75:25 | 1.7/1.9 | 1.7/1.9 | 1.7/1.9 | |

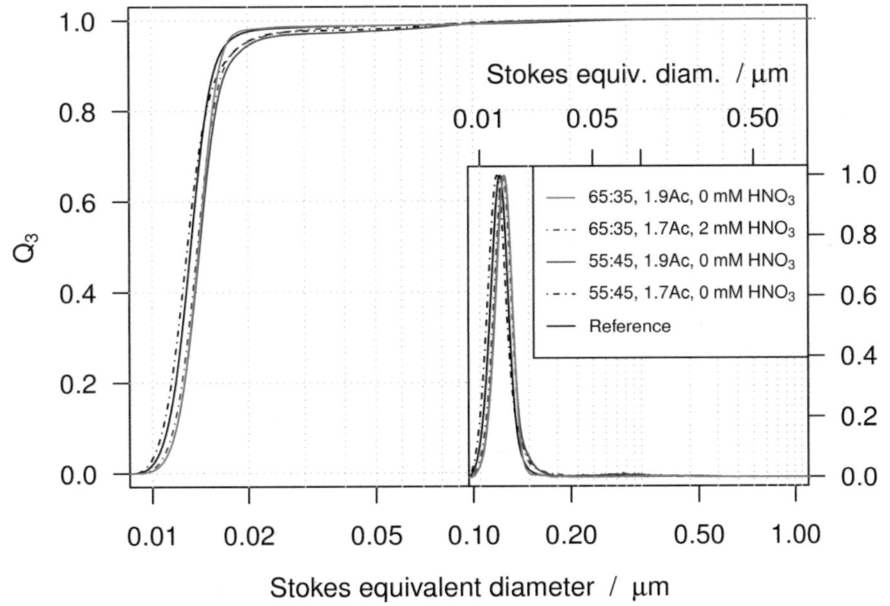


**Figure 6.16:** Particle size distribution of dispersions synthesized from mixtures of iron(II) acetate and iron(III) nitrate with narrow particle size distributions similar to the distributions of the highly transparent dispersions shown in Figure 6.13 (Reference). For all samples: $\dot{m}_{total}$ = 6.07 kg/h, $\phi_m$ = 0.2, and $c_{\text{product Fe}}$ = 0.001 mol/L.

experiments: 13.2 nm $\leq$ mode < 15.8 nm) and overall shape of the distributions. Although some of the samples showed a fraction of particles > 20 nm and/or a tailing of the main peak in the direction of larger particles, only marginal differences in the shape of the main peak were observed.

In order to explore the reproducibility of these results, replicate runs were performed for some of the compositions. Representative results are given for one composition in Table 6.7. No significant variations were observed between the runs for the mode ($x_{k,3}$), the median ($x_{50,3}$), and $x_{10,3}$ (characterizes the left side of the particle size distribution). However, differences in the tailing of the main peak were observed between the runs. In particular, one experimental run (Run C) showed a stronger tailing leading to significantly higher values of $x_{90,3}$, $x_{g,3}$ and $\sigma_{g,3}$.

**Table 6.7:** Characteristic values of the particle size distributions synthesized using a mixture of iron(II) acetate (65%) and iron(III) nitrate (35%) with the addition of 0.003 mol/L of acetic acid (acetate factor = 1.9) at $\dot{m}_{total}$ = 6.07 kg/h, $\phi_m$ = 0.2, and $c_{\text{product Fe}}$ = 0.001 mol/L, in comparison with the Reference sample (Ref., cf. Figure 6.16). An overlay of the corresponding particle size distributions is shown in Figure E.16 on page 250.

| Label | $pH_{salt}$/ $pH_{product}$ | $x_{k,3}$ nm | $x_{50,3}$ nm | $x_{10,3}$ nm | $x_{90,3}$ nm | $x_{g,3}$ nm | $\sigma_{g,3}$ |
|---|---|---|---|---|---|---|---|
| Run A | 3.7/3.1 | $14.0 \pm 0.2$ | $13.9 \pm 0.2$ | $11.8 \pm 0.2$ | $16.2 \pm 4.0$ | $14.3 \pm 0.2$ | $1.37 \pm 0.09$ |
| Run B | 3.8/3.1 | $13.7 \pm 0.2$ | $13.9 \pm 0.2$ | $11.6 \pm 0.2$ | $17.2 \pm 4.0$ | $14.6 \pm 0.2$ | $1.38 \pm 0.09$ |
| Run C | 3.8/3.1 | $13.7 \pm 0.2$ | $14.0 \pm 0.2$ | $11.6 \pm 0.2$ | $25.6 \pm 4.0$ | $16.0 \pm 0.2$ | $1.63 \pm 0.09$ |
| Ref. | 3.0/3.5 | $13.3 \pm 0.3$ | $13.1 \pm 0.3$ | $11.0 \pm 0.2$ | $15.5 \pm 0.6$ | $13.5 \pm 0.4$ | $1.36 \pm 0.06$ |

Considering the high reproducibility of the position and overall shape of the main peak, it is possible to discern that the value of the mode is affected by changes in the composition of the starting solution (acetate factor, ratio of Fe(II) to Fe(III), and addition of $HNO_3$) and that the effect of changing one factor depends on the values of the other two. This is most clearly seen for the case of increasing the addition of $HNO_3$ at different values of the ratio of Fe(II) to Fe(III) for a given acetate ratio, as shown in Table 6.8 (and graphically in Figure E.17 – E.19 on page 252 – 254) . Due to its higher variability, it is not possible to discern such effects for the geometric standard deviation of the distributions (for data on pH values and $\sigma_{g,3}$ of educt and product, please consult the Table E.13 and E.14 on page 251 and 251, respectively.

**Table 6.8:** Mode ($x_{k,3}$, ±0.2 nm) of the particle size distribution (obtained by differential sedimentation analysis) from the dispersions synthesized from different compositions using mixtures of iron(II) acetate and iron(III) nitrate at $\dot{m}_{total}$ = 6.07 kg/h, $\phi_m$ = 0.2, and $c_{\text{product Fe}}$ = 0.001 mol/L.

| | | Addition of $HNO_3$ | | |
|---|---|---|---|---|
| Acetate factor | Fe(II):Fe(III) | 0 mmol/L | 2 mmol/L | 4 mmol/L |
| 1.7 | 75:25 | 15.3 | 15.8 | 15.3 |
| 1.7 | 65:35 | 13.7 | 14.0 | 15.4 |
| 1.7 | 55:45 | 13.2 | 14.8 | 14.8 |
| 1.9 | 75:25 | 15.8 | 14.5 | 14.1 |
| 1.9 | 65:35 | 14.0 | 14.6 | 14.9 |
| 1.9 | 55:45 | 13.5 | 14.4 | 15.0 |
| 2.1 | 55:45 | 14.1 | 14.8 | 15.6 |
| 2.1 | 45:55 | 14.3 | 14.8 | 15.1 |

To further explore whether it is possible to tune the particle size distribution with respect to the position of the main peak by variation of the starting composition, the range of $HNO_3$ addition was extended for selected compositions of the starting solutions (ratio of Fe(II) to Fe(III) and acetate factor) for which the effect of adding $HNO_3$ on the mode of the particle size distribution was comparatively strong. Table 6.9 summarizes the progression (generally increasing values from 14.1 to 16.2 nm) of the mode upon increasing the nitric acid content for a mixture of iron(II) acetate (55%) and iron(III) nitrate (45%) and an acetate factor of 2.1. One representative particle size distribution of each sample (as measured with a disc centrifuge) is shown in Figure E.20 on page 255. There are several possible reasons for the observed increase in the mode of the particles size distributions. First, it is possible that the size of the primary particles increased. Second, the increase may be due to an increase in the size of the agglomerates/aggregates. Finally, it is also possible that differences in the density of the particles due to structural differences in the nanoregime lead to increased or decreased measured particle sizes. Unfortunately, it was not possible to determine the contribution of each of these possibilities to the observed effect within this work. Nevertheless, it was possible to obtain some insights with the help of transmission electron micrographs and Raman spectroscopy of a sample synthesized from a mixture of iron(II) acetate (65%) and iron(III) nitrate (35%) with an acetate factor of 1.9.

**Table 6.9:** Mode ($x_{k,3}$) of the particle size distribution as well as pH-values of salt solution and product of iron oxide nanoparticle dispersions synthesized with $\phi_m$ = 0.2 and $\dot{m}_{total}$ = 6.07 kg/h using a mixture of iron(II) acetate (55%) and iron(III) nitrate (45%) with the addition of 0.003 mol/L of acetic acid (acetate factor = 2.1) and $c_{\text{product Fe}}$ = 0.001 mol/L for different amounts of added nitric acid.

| Addition of $HNO_3$ / mmol $L^{-1}$ | 0 | 2 | 4 | 6 | 8 | 10 |
|---|---|---|---|---|---|---|
| $x_{k,3}$ / nm (± 0.2 nm) | 14.1 | 14.8 | 15.6 | 15.3 | 15.9 | 16.2 |
| $pH_{salt}$ | 3.1 | 2.8 | 2.4 | 2.3 | 2.1 | 2.0 |
| $pH_{product}$ | 2.9 | 2.7 | 2.7 | 2.6 | 2.5 | 2.4 |

Figure 6.17 shows representative overview and detailed TEM images of the selected sample. The overview image strongly resembles TEM images of particles synthesized from aqueous iron(II) acetate solutions (highly transparent, shown in Figure 6.14). The detailed micrograph almost exclusively shows fairly spherical particles around 6 nm in the size range from 4.5 to 7.5 nm. Analyses of the sizes of the primary particles revealed larger primary particles (by around 1 nm) in comparison to the primary particles obtained from the initial acidified iron(II) acetate solutions (see Figure 6.14 and discussion thereof). However, due to the difficulties in precisely measuring the size of particles in the TEM images, not sufficient and reliable data could be obtained to test the statistical significance of the observed difference. A similar tendency is also observed for the agglomerate size distribution (see Table 6.7).

To clarify whether there are also similarities in the strucutre of the particles, we resorted to Raman spectroscopy. Raman spectra of the concentrated dispersions were taken at different laser power settings, as well as spectra from the same sample upon air drying. Representative spectra are displayed in Figure 6.18, which also contains the Raman spectra of ferrihydrite with marked peak positions for reference purposes. The Raman spectra of the dispersed particles show the characteristic asymmetric ferrihydrite main band at around 710 $cm^{-1}$. The band maximum is slightly shifted towards higher wavenumbers compared to the freeze-dried ferrihydrite reference sample. The two broad bands at around 350 $cm^{-1}$ and 510 $cm^{-1}$ are also visible with slight shifts compared to the reference sample but within the range as it is observed for ferrihydrite samples in case of both power settings. These shifts are very similar to those detected investigating particles synthesized with acidified iron acetate (cf. Figure 6.15). No significant changes were noticed in the Raman spectra upon increasing the laser power setting. Thus, the spectra of the nanoparticle dispersions suggest that ferrihydrite is the dominant iron oxide phase the nanoparticles consist of. However, the presence of maghemite fractions (see discussion above) can not be ruled out.

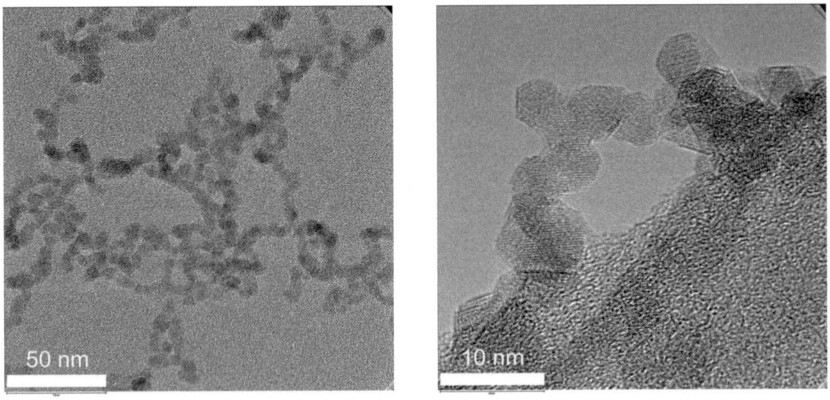


**Figure 6.17:** TEM images of particles from iron oxide nanoparticle dispersions synthesized with $\phi_m = 0.2$ and $\dot{m}_{total} = 6.07$ kg/h using a mixture of iron(II) acetate (65%) and iron(III) nitrate (35%) with the addition of 0.003 mol/L of acetic acid (acetate factor = 1.9) and $c_{\text{product Fe}} = 0.001$ mol/L.

The overall shape of the Raman spectrum was conserved upon air-drying of the sample. However, the band in the spectral range from 1300 $cm^{-1}$ until 1400 $cm^{-1}$ is slightly shifted towards higher wavenumbers as well as the band at 710 $cm^{-1}$ is not as asymmetric and has not an as steep rise of the left edge compared to the reference sample and the sample measured as dispersion. This may indicate the formation (or increase of the fraction) of maghemite, into which ferrihydrite can be transformed before transformation into hematite (Mazzetti and Thistlethwaite, 2002). Nevertheless, the structure of these sample appears much more stable upon drying in air compared to the sample that was only synthesized from iron acetate (see Figure 6.15). Here, it is interesting to note, that the position of the main band is the same in the air-dried state as in the dispersed state and no shift to slightly lower wavenumbers as observed for the freeze-dried reference sample(s) occured. However, the unidentified band at 914 $cm^{-1}$ already seen in spectra of Figure 6.15 also appears in spectra of samples measured as dispersion (Figure 6.18). This mode is absent in case of dried sample(s) as discussed previously. The possibility of structural transformations upon (freeze-)drying was stated by Peak and Regier (2012).

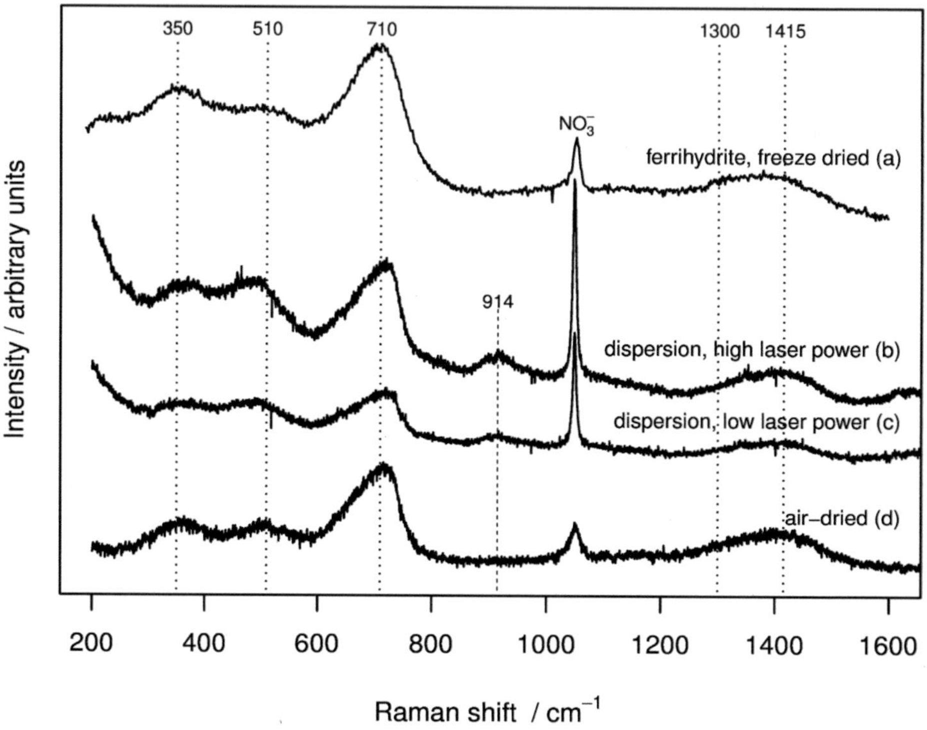


**Figure 6.18:** Raman spectra from: a) reference sample 2-line ferrihydrite according to Schwertmann and Cornell (2000); b) iron oxide nanoparticle dispersion synthesized with $\phi_m = 0.2$ and $\dot{m}_{total} = 6.07$ kg/h using a mixture of iron(II) acetate (65%) and iron(III) nitrate (35%) with the addition of 0.003 mol/L of acetic acid and $c_{\mathrm{product\,Fe}} = 0.001$ mol/L, measured as a concentrated dispersion ('hanging drop') with a laser power of 2.0 mW; c) same sample as b) but measured with a laser power of 0.5 mW; d) same sample as b) but measured after air-drying. The spectra are shifted vertically for clarity.

## 6.3 Interim summary

Electrostatically stabilized iron oxide nanoparticle dispersions were prepared from iron(II) acetate, iron(III) nitrate, and mixtures of both with or without the addition of the respective acids. The dispersions were characterized by differential centrifugal sedimentation analysis—a technique both powerful and convenient for obtaining particle/agglomerate size distributions—and by Raman spectroscopy using a modified setup that allows precise measurements directly in concentrated dispersions.

Using these techniques, it was found that dispersions synthesized from pure aqueous iron(III) nitrate solutions exhibit a bimodal distribution regarding both primary particles as well as agglomerates. However, upon increasing the total mass flow rate and mixing ratio between the aqueous metal salt stream and supercritical water stream, the fraction of the larger particles and agglomerates (around 30 nm and 65 nm, respectively) was decreased in favor of the smaller particles and agglomerates (around 7 nm and 20 nm, respectively) up to a point were essentially no larger particles were present. Moreover, a mixture of $\alpha$-$Fe_2O_3$ and $\gamma$-$Fe_2O_3$ and/or ferrihydrite were detectable when investigating the dispersions by means of Raman spectroscopy. Furthermore, combining analyses of the dispersions with analyses of the particles as dry powders, it was shown that the structure and size of the synthesized particles can change upon drying with the potential of misleading results when measuring the particles only as dried powders. However, further Raman spectroscopic investigations of iron oxides in dispersions are needed to discriminate and obtain more information regarding structure and size of the particles in dispersed form. Such insights are of interest not only for hydrothermal synthesis due to the overall importance of iron oxides/oxyhydroxides.

Significantly narrower particle/agglomerate size distributions with primary particles most likely of uniform structure were obtained if mixtures of iron(II) acetate and iron(III) nitrate were used as a starting material instead of pure iron(III) nitrate under otherwise same flow conditions (see Figure 6.19). In this study, the potential of varying the primary particle and agglomerate size distribution continuously through changing the composition of the starting solutions was shown. However, further studies are necessary to discriminate between the different effects leading to a shift of the particle size distributions as observed with differential centrifugal sedimentation. Nevertheless, the approach seems reasonable to screen for favorable compositions of starting material as an additional variable in optimizing the hydrothermal synthesis of metal oxide nanoparticles. This argument is strengthened by the finding that the composition of the educt solutions also has an influence on the sensitivity of the product properties to fluctuations in the process/flow conditions—an important aspect for robust up-scaling.

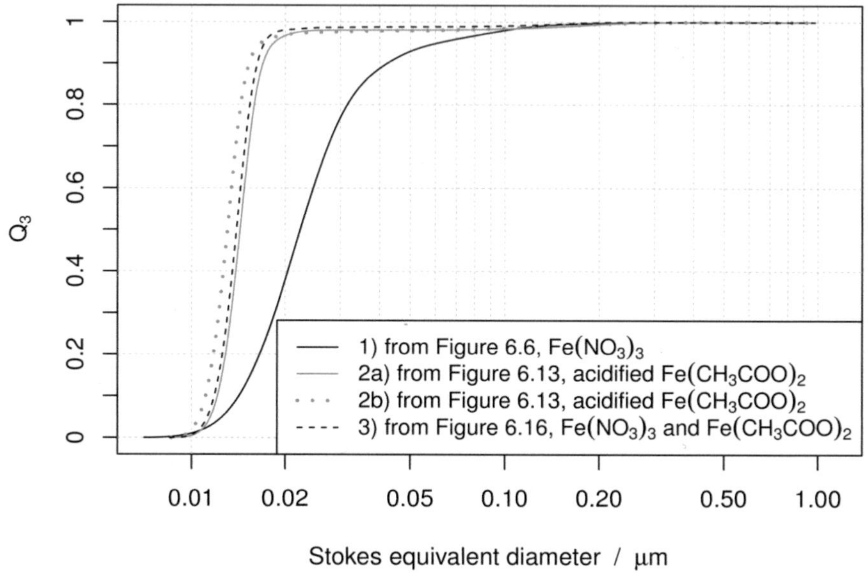


**Figure 6.19:** Particle size distributions for selected dispersions prepared from different starting materials under the same flow conditions ($\phi_m = 0.2$ and $\dot{m}_{total} = 6.07$ kg/h). 1) $c_{\text{product Fe}} = 0.0055$ mol/L; 2a) $c_{\text{product Fe}} = 0.0055$ mol/L, 5 mmol/L $HNO_3$ added to the starting solution; 2b) $c_{\text{product Fe}} = 0.001$ mol/L, 2 mmol/L $HNO_3$ added to the starting solution; 3) $c_{\text{product Fe}} = 0.001$ mol/L, Fe(II):Fe(III) 65:35, 1.9 Ac, no added $HNO_3$.

The use of a highly sensitive material system such as iron oxides/oxyhydroxides together with methods for analyzing the particles/particle agglomerates directly in the dispersions provides a good basis, e.g., for studies of mixing in continuous hydrothermal synthesis, allowing both the comparison of different mixing geometries as well as the optimization of the processs regarding reproducibility. Future studies could combine the methods described here with ongoing research on optimizing the mixing process in order to further narrow the particle size distributions and increase the concentration of metal oxides in stable product dispersions.

# Chapter 7

# In situ functionalization of iron oxide nanoparticles

The extension of the continuous hydrothermal synthesis to produce functionalized metal oxide nanoparticles with terminal groups suitable for further modification by *click* chemistry is reported in this chapter. By definition, *click* reactions are simple, efficient and applicable in diverse modular approaches without the formation of significant side products (Kolb et al., 2001), making *click* chemistry a promising tool in colloidal science (Lutz and Zarafshani, 2008). The use of molecules containing a terminal alkyne group allows straightforward coupling with azide-terminated molecules via the most used and well-known click reaction, the copper(I)-catalyzed azide-alkyne cycloaddition (CuAAC) (Such et al., 2012). Versatile and facile methods for the preparation of azide-terminated polymers and other molecules are described in the literature (Lu et al., 2009). Magnetic iron oxide nanoparticles served as a model system because they are useful for a broad range of applications in the biomedical, biotechnological and environmental field on account of their biocompatibility and low toxicity (Gupta and Gupta, 2005; Zhou et al., 2011). In the explorative study reported in this chapter, the effect of different process variables—namely the use of different mixing setups with varying temperature, iron salts, and organic modifiers—on the properties of the resulting iron oxide nanoparticle dispersions is investigated. There are reasons to expect that changes in these variables will result in measurable changes in the product dispersions, namely:

- It is known that the use of different iron salts can lead to different structures of the iron oxide nanoparticles. In turn, this can influence the reactivity of the particle surface—an important aspect for in situ functionalization.
- Given that organic capping agents affect particle formation, growth and stability, it appears reasonable that the chain length of the modifiers—which in turn

affects e.g. diffusion time to the surface and the ability to stabilize the particles via steric interactions—may noticeably affect the properties of the resulting dispersions.

- In terms of particle formation and growth, it could make a substantial difference whether the organic modifiers are already present when the compressed cold iron salt solution is rapidly heated to the reaction conditions or the organic modifiers are added afterwards.
- The reaction temperature not only affects particle formation and growth via the properties of the near-critical or supercritical water but can also influence the reaction medium through the partial thermal decomposition of the added organic modifiers. For this reason, the stability of the added carboxylic acids under near-critical and supercritical conditions was also studied in the absence of the iron salts.

A summary of the experimental runs performed is given below (Table 7.1). Refer to Chapter 3 for the different configurations and mode of operation of the pilot plant used within this study.

**Table 7.1:** Summary of the experimental runs to synthesize primary functionalized iron oxide nanoparticles using iron(III) nitrate and ammonium iron(III) citrate as iron salts, hexanoic acid (HexaAc), 5-hexynoic acid (5-HexyAc), and 10-undecynoic acid (10-UndecyAc) as organic modifiers and different mixing setups (cf. Figure 4.6 on page 85).

| Run | mixing setup | iron salt | $T_{set}$ [K] | $T_w$ [K] | carboxylic acid | org : salt [mol : mol] |
|---|---|---|---|---|---|---|
| 1A | "salt-organic" | nitrate | 673 | 707 | - | - |
| 1B | "salt-organic" | nitrate | 673 | 707 | HexaAc | 20 : 1 |
| 1C | "salt-organic" | nitrate | 673 | 707 | HexaAc / 5-HexyAc | 20 : 1 |
| 1D | "salt-organic" | nitrate | 673 | 707 | 5-HexyAc | 20 : 1 |
| 2A | "salt-organic" | nitrate | 573 | 611 | 5-HexyAc | 15 : 1 |
| 2B | "salt-organic" | nitrate | 623 | 666 | 5-HexyAc | 15 : 1 |
| 2C | "salt-organic" | nitrate | 673 | 703 | 5-HexyAc | 15 : 1 |
| 3A | "salt-organic" | citrate | 573 | 623 | 5-HexyAc | 15 : 1 |
| 3B | "salt-organic" | citrate | 623 | 667 | 5-HexyAc | 15 : 1 |
| 3C | "salt-organic" | citrate | 673 | 703 | 5-HexyAc | 15 : 1 |

*Continued on next page*

*Continued from previous page* **(Table 7.1)**

| Run | mixing setup | iron salt | $T_{set}$ [K] | $T_w$ [K] | carboxylic acid | org : salt |
|---|---|---|---|---|---|---|
| 4A | "organic-salt" | nitrate | 573 | 626 | 5-HexyAc | 15 : 1 |
| 4B | "organic-salt" | nitrate | 623 | 668 | 5-HexyAc | 15 : 1 |
| 4C | "organic-salt" | nitrate | 673 | 701 | 5-HexyAc | 15 : 1 |
| 5A | "organic-salt" | citrate | 573 | 625 | 5-HexyAc | 15 : 1 |
| 5B | "organic-salt" | citrate | 623 | 665 | 5-HexyAc | 15 : 1 |
| 5C | "organic-salt" | citrate | 673 | 707 | 5-HexyAc | 15 : 1 |
| 6A | "mixed" | citrate | 573 | 625 | 5-HexyAc | 15 : 1 |
| 6B | "mixed" | citrate | 623 | 667 | 5-HexyAc | 15 : 1 |
| 6C | "mixed" | citrate | 673 | 707 | 5-HexyAc | 15 : 1 |
| 7A | "undecynoic" | nitrate | 673 | 639 | 10-UndecyAc | 10 : 1 |
| 7B | "undecynoic" | citrate | 673 | 677 | 10-UndecyAc | 10 : 1 |
| 7C | "undecynoic" | nitrate | 673 | 734 | 10-UndecyAc | 10 : 1 |
| 7D | "undecynoic" | citrate | 673 | 734 | 10-UndecyAc | 10 : 1 |

## 7.1 Optical examination of the non-functionalized and functionalized dispersions

At first, a series of experiments was conducted (Run 1A-D) using iron nitrate as the iron salt, and starting with the synthesis of unfunctionalized iron oxide particles (Run 1A). In the next run (1B), a typical monofunctional surface modifier bearing one carboxyl group—namely hexanoic acid—was used for functionalization of the particles. In a subsequent run (1C), an equimolar mixture of hexanoic and 5-hexynoic acid—a derivative of hexanoic acid bearing an alkyne group at the opposite end of the chain and thus suitable for *click* chemistry—was used for functionalization. For the last run of that series (Run 1D), only 5-hexynoic acid was used. Dispersions of these experimental runs are shown in Figure 7.1.

Highly transparent and stable dispersions of a reddish color were obtained for the unfunctionalized case. A detailed characterization of iron oxide dispersions synthesized from iron nitrate was described in the previous chapter (6). In case of using hexanoic acid, an emulsion was obtained of droplets with a dark brown-reddish color in a yellowish continuous phase, whereas the droplets started to coalesce and to build a separate phase on top of the remaining aqueous phase (see Figure 7.1b encircled in white). Due to the very low solubility of the hexanoic acid in water at low pH-value an organic phase is formed at room temperature. As seen from the color of the dispersion, the synthesized particles preferably remained in the organic phase which is an indication for the successful functionalization of the particles with hexanoic acid. When a mixture of hexanoic and 5-hexynoic acid was used, an unstable dispersion of brownish color was formed from which particles started to sediment quickly (noticable by the established color gradient in the vicinity of the air/water interface in Figure 7.1c) with a small layer of foam formed at the top. When 5-hexynoic acid was used alone, the dispersion appeared darker and slightly more foam was formed. In addition, the sedimentation of the particles was slower.

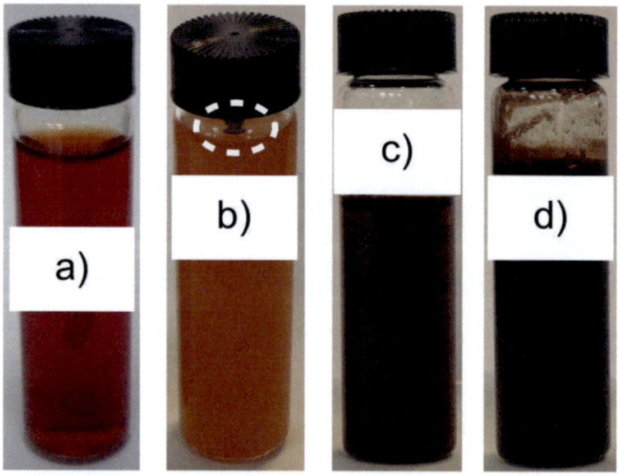


**Figure 7.1:** Product dispersions from: a) Run 1A, b) Run 1B, c) Run 1C, d) Run 1D. The labels with the white boxes were introduced on top of the original sample labels for the sake of clarity.

For a rapid prescreening for successful functionalization, a phase transfer test was always carried out with the as obtained dispersions by mixing with toluene. The particles synthesized in the presence of hexanoic and 5-hexynoic acid (Run 1C), which sedimented quickly in the aqueous phase, transferred into the organic phase as shown in Figure 7.2a. A fraction of the particles was well dispersed, whereas another fraction of highly agglomerated particles was found in the vicinity of the interface of the two phases. Concerning the functionalization when 5-hexynoic acid was used alone, depending on the specific conditions the extent of phase transfer ranged from a small fraction of particles that transferred into the organic phase up to practically complete phase transfer. However, a fraction of these transferred particles formed highly agglomerated particles which sedimented through the aqueous phase as shown in Figure 7.2c.

The functionalization with 5-hexynoic acid was systematically explored by varying the reaction temperature, the identity of the iron salt and the mixing setup. For experimental runs carried out using iron nitrate as starting material, the particles preferentially transferred into the organic phase for all temperatures and mixing setups ("salt-organic" and "organic-salt") investigated. However due to the formation of dense agglomerates, fractions of the particles always sedimented—also through the aqueous phase - as already described above. An example is given by Figure 7.2b (Run 2B). When ammonium iron citrate was used as starting material, particles preferably transferred into the organic phase only when a reaction temperature of 623 K or higher was set.

In Figure 7.2c and 7.2d, product dispersions are shown that were obtained for a reaction temperature of 673 K using the setup "organic-salt" and "mixed" respectively (Run 5C and Run 6C) . In both cases, it is obvious that a large fraction of particles form densely packed particle agglomerates when dispersed in the toluene phase. These agglomerates are first trapped in the phase boundary and then sediment through the aqueous phase.

**Figure 7.2:** Phase transfer test of selected particles synthesized in Run: a) 1C (in the presence of a mixture of hexanoic and 5-hexynoic acid), b-d) 2B, 5C and 6C (5-hexynoic acid), e) - f) 7A and 7D (10-undecynoic acid). Upper phase: rich in toluene, lower phase: aqueous. The labels with the white boxes were introduced on top of the original sample labels for the sake of clarity.

Particles synthesized from iron nitrate in the presence of 10-undecynoic acid mostly transferred into the organic phase when a reaction temperature of 573 K was set and formed a fairly stable dispersion in toluene as shown in Figure 7.2 e). For reaction temperatures of 623 K and 673 K, a preference of the residence of the particles could not be clearly observed. However, a fraction of the particles was preferably at the interface. Due to the addition of the ammonia solution before the reaction stream was cooled, unreacted dissociated 10-undecynoic acid could form a bilayer or mutliple layers with the 10-undecynoic acid molecules that attached to the surface of the particles leading to an electrosteric stabilization mechanism (see also page 155). When iron citrate was used as the iron salt, the particles transferred into the organic phase and formed a fairly stable and homogenous dispersion (see Figure 7.2 f). The turbidity of the aqueous phase is most likely caused by unreacted 10-undecynoic acid not fully dissociated due to an insufficient amount of added ammonia solution.

## 7.2 Stability of alkynoic acids at near- and supercritical conditions

Due to the harsh conditions for the alkynoic acids in the near- and supercritical environment, an important question to answer was whether the molecules for functionalization were sufficiently stable. The occurence of foam already indicated that at least a fraction of the alkynoic acids could decompose and release gaseous species. The stability of 5-hexynoic acid and 10-undecynoic acid was investigated at process conditions by determining the concentration of the two alkynoic acids in the product stream and setting this in relation to the expected concentration in that stream in the case of no degradation. The results are shown in Figure 7.3. When the stability of 5-hexynoic was tested at the lowest synthesis temperature (573 K), about 90% was recovered in the product stream. As expected, with increasing temperature the remaining fraction decreased to about 70% at the highest reaction temperature (698 K). The temperature of the reaction zone was increased further because the alkynoic acids shortly pass through conditions of higher temperature when the mixing setup "organic-salt" is in use. With temperatures higher than 698 K, the remaining fraction decreased further but not significantly. This indicates that no strong differences should exist between available molecules for functionalization for the different mixing setups used. However, the actual measured values can only count as a rough estimate due to differences in the pH-value and the presence of other chemical species under particle synthesis conditions. There is also the possibility that, upon the attachment of the alkynoic acids to the surface, these molcules are stabilized with respect to decarboxylation, a possible and likely decomposition pathway (Belsky et al., 1999; Takami et al., 2008).

When the stability of 10-undecynoic acid was investigated under process conditions, the carboxylic acid—like in the experiments for the in situ functionalization (experimental series 7)—was dissolved in ethanol to be delivered as a fluid (melting point of 10-undecynoic acid 315 K) into the pilot plant. Therefore, differences in the solvent and reaction media between the stability test of 5-hexynoic acid and 10-undecynoic acid exist. However, as expected due to the longer chain length, the remaining fraction of 10-undecynoic acid was significantly lower for all reaction temperatures. At the lowest and highest reaction temperatures investigated (573 K and 673 K), around 90% and 50%, respectively, were recovered in the reaction stream. The difference between the recovered fractions of 5-hexynoic acid and 10-undecynoic acid was for all reaction temperatures about 0.3. It has to be taken into acount that also at the lowest remaining fraction detected, sufficient carboxylic acid is available to fully cover the surface of nanoparticles even for sizes below 10 nm since the carboxylic acids are provided in large excess for the functionalization experiments.

To further rule out the possibility that the alkynoic acids decompose fully or at least at a much higher rate when iron species are present, the remaining fraction of the respective carboxylic acids in solution were determined for selected product dispersions. Here, of course, carboxylic acids that are in any way associated with the particles (possibilities are: covalently bound, chemisorbed, physisorbed, and molecules that are interacting with the chains that are associated with the particle) are not detected via this method. Nevertheless, the remaining fractions of alkynoic acids were in the range of 10–50% when iron nitrate or iron citrate was used as starting material for the nanoparticles (data not shown), indicating that sufficient organic material is present for functionalization also in the presence of iron species.

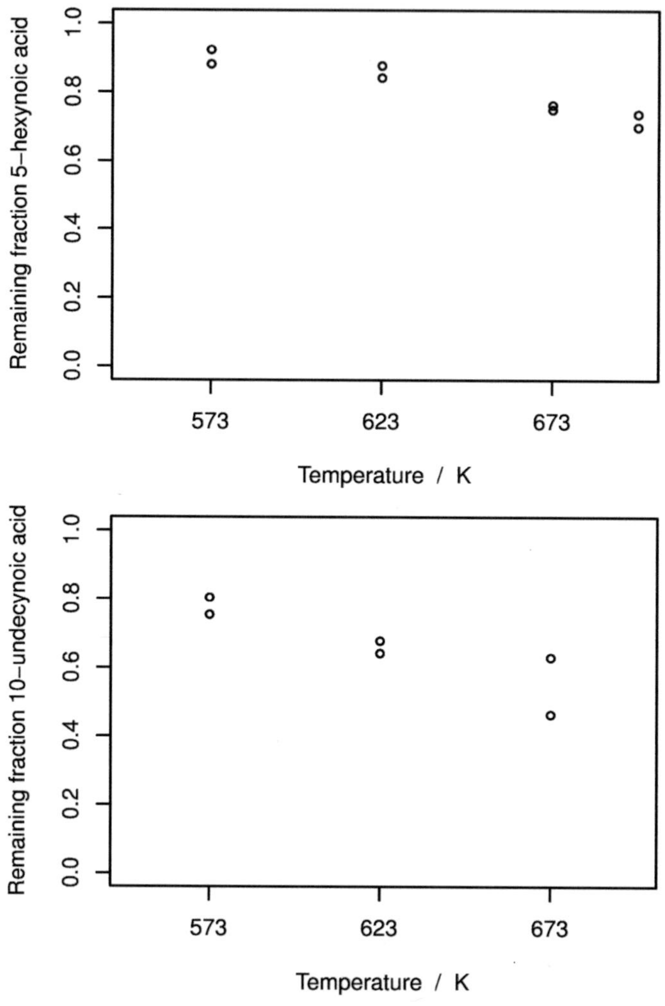


**Figure 7.3:** Remaining 5-hexynoic and 10-undecynoic acid under different temperatures in the reaction zone ($p = 30$ MPa). The analytical error was estimated to be 0.04 (one standard deviation). For each reaction temperature, two samples were taken at different times after reaching steady state.

# 7.3 Size, morphology and crystal structure of the particles

The different colors of the dispersions shown in Figure 7.1 and Figure 7.2, from orange-red over brown to black, already point to differences in the structure of the synthesized iron oxide nanoparticles, namely hematite, maghemite and magnetite (Cornell and Schwertmann, 2003). The structure, as well as the mean crystallite sizes were investigated in detail for selected samples by means of powder X-ray diffraction (PXRD). By applying quantitative Rietveld analysis on the basis of structure data reported in literature (Finger and Hazen, 1980; Haavik et al., 2000; Pecharromán et al., 1995), the compositions of the iron oxide phases were calculated. Because of the strong similarities of magnetite ($Fe_3O_4$) and maghemite ($\gamma$-$Fe_2O_3$) in the diffraction patterns (see Figure 7.4), especially for nanoscaled crystallites, they were considered together in this analysis.

When analyzing the selected samples, three different types of diffraction patterns were observed. An example for each type is given in Figure 7.4:

- The diffraction peaks could be attributed to one iron oxide phase and were fitted to calculate the volume-weighted mean crystallite size which represents the nanocrystallites reasonably (top diffraction pattern, from experimental run 5C)
- The diffraction pattern contained peaks from hematite and magnetite/maghemite (central diffraction pattern, from experimental run 4B). Due to the large difference of the crystallite sizes of the two phases, the peaks from the iron oxide phases present in form of very small crystallites (very broad peaks) are barely visible.
- The observed diffraction pattern was clearly a superposition of at least two fractions of particles with fairly different crystallite sizes (bottom diffraction pattern, from experimental run 5B) of the same iron oxide phase. In these cases the diffraction pattern was fitted for a monomodal as well as for a bimodal distribution. An example of this type is shown in Figure 7.5.

Both hematite and magnetite/maghemite were detected for all the particles synthesized using iron nitrate as starting material (second type of diffraction patterns). When iron citrate was used only magnetite/maghemite was detected. Regarding the particle size distribution, either the diffraction pattern was reasonably fitted assuming a monomodal distribution or a bimodal fit was clearly necessary.

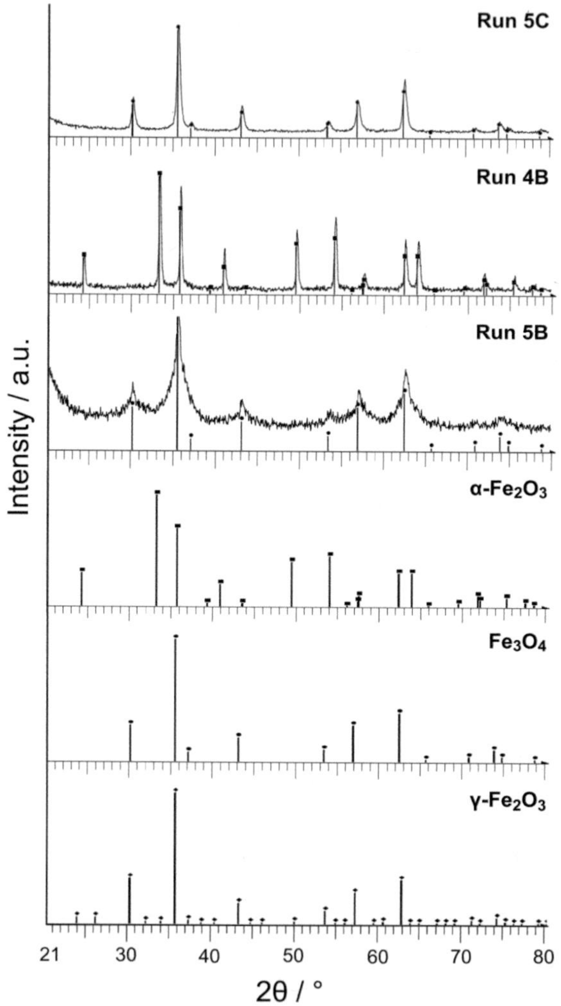


**Figure 7.4:** XRD patterns of particles of Run 5C, 4B and 5B; positions and relative intensities of the lines for the relevant iron oxide phases are shown at the bottom (powder diffraction file (PDF) 33-664 for $\alpha$-$Fe_2O_3$, PDF 19-629 for $Fe_3O_4$, PDF 39-1346 for $\gamma$-$Fe_2O_3$ from the International Centre for Diffraction Data, ICDD, USA).

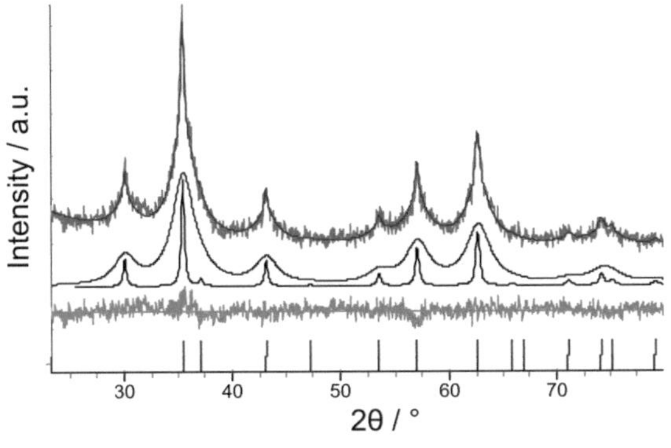


**Figure 7.5:** XRD pattern of particles of Run 5B, top: raw data with whole pattern fit, middle: bimodal fit, bottom: residual of bimodal fit and raw data.

The results of the crystallite size and composition analysis by means of PXRD are given in Table 7.2. For the first series of in situ functionalized particles with changing composition of the carboxylic acids using the "organic-salt" mixing setup, the fraction of hematite decreased from about 50% when only hexanoic acid (run 1B) was used via about 40% for the 1:1 mixture of hexanoic and 5-hexynoic acid (run 1C) to about 10% when only 5-hexanoic acid (run 1D) was used. Upon this change in composition, the size of the hematite crystallites decreased between run 1B and 1C from around 58 nm to 47 nm with no further significant decrease for run 1C while the mean size of the maghemite/magnetite crystals increased from 3.4 nm up to 8.3 nm. The differences observed in the composition of the iron oxides could be linked to differences in the partial decomposition of the organic modifiers. This possibility arises from the sensitivity of the iron oxide material system—which shows varied physical, chemical, and structural properties with continuous or sudden change between—and the observation that the structure and properties of iron oxide nanoparticles depend on their history of formation and surrounding medium (Chernyshova et al., 2007). In particular, it has been proposed that the partial thermal decomposition of ammonium citrate leads to reduction of $Fe^{3+}$ to $Fe^{2+}$ to account for the formation of $Fe_3O_4$ Adschiri et al. (1992). This observation implies that the reaction environment can be altered not only by the counterion of the iron salt (nitrate vs. ammonium citrate) but also by the added organic modifiers. A similar trend was observed when the reaction temperature was increased from 623 K to 673 K for the mixing setup "salt-organic". The fraction and mean crystallite size of hematite decreased from about 85% to about

**Table 7.2:** PXRD results of selected samples of functionalized particles with 5-hexynoic and 10-undecynoic acid. PSD = Particle Size Distribution (assumption of a monomodal or bimodal distribution concerning the maghemite/magnetite phase); m = monomodal distribution; b = bimodal distribution.

| | Experimental conditions | | | Crystallite size [nm] | | Composition [%] | | |
|---|---|---|---|---|---|---|---|---|
| Run | mixing setup | iron salt | $T_{reactor}$ [K] | $\alpha$-$Fe_2O_3$ | $\gamma$-$Fe_2O_3$ or $Fe_3O_4$ | $\alpha$-$Fe_2O_3$ | $\gamma$-$Fe_2O_3$ or $Fe_3O_4$ | PSD |
| 1B | "organic-salt" | nitrate | 673 | 57.8 | 3.4 | 52.8 | 47.2 | m |
| 1C | "organic-salt" | nitrate | 673 | 47.2 | 7.6 | 39.4 | 60.6 | m |
| 1D | "organic-salt" | nitrate | 673 | 46.5 | 8.3 | 10.5 | 89.5 | m |
| 2B | "salt-organic" | nitrate | 623 | 92.8 | 3.4 | 85.5 | 14,5 | m |
| 2C | "salt-organic" | nitrate | 673 | 71.1 | 6.6 | 68.3 | 31.7 | m |
| 3B | "salt-organic" | citrate | 623 | - | 4.9 | - | 100 | m |
| | | | | | 4.3/62 | | 90.7/9.3 | b |
| 3C | "salt-organic" | citrate | 673 | - | 33.2 | - | 100 | m |
| 4A | "organic-salt" | nitrate | 573 | 5.6 | 1.9 | 86.5 | 13.5 | m |
| 4B | "organic-salt" | nitrate | 623 | 82.3 | 3.9 | 81.9 | 18.1 | m |
| 5B | "organic-salt" | citrate | 623 | - | 5.7 | - | 100 | m |
| | | | | | 4.0/38.3 | | 88.0/12.0 | b |
| 5C | "organic-salt" | citrate | 673 | - | 26.7 | - | 100 | m |
| 6B | "mixed" | citrate | 623 | - | 6.7 | - | 100 | m |
| | | | | | 4.5/38.2 | | 86.2/13.8 | b |
| 6C | "mixed" | citrate | 673 | - | 30.8 | - | 100 | m |
| 7C | "undecynoic" | nitrate | 673 | 49.7 | 5.2 | 25.3 | 74.7 | m |
| 7D | "undecynoic" | citrate | 673 | | 21.6 | 2.2 | 97.8 | m |

70% and from around 93 nm to 71 nm, respectively, whereas the crystallite size of maghemite/magnetite increased from 3.4 to 6.6 nm. Fairly similar results (with respect to the compositions and sizes of the crystallites) were obtained for the mixing setup "organic-salt" at a reaction temperature of 623 K. However, the larger particles of hematite structure were about 10 nm smaller (a decrease in the mean crystallite size of slightly over 10%) whereas the smaller particles of maghemite/magnetite structure were slightly bigger. A similar trend was observed between particles synthesized at the two mixing setups at a reaction temperature of 673 K. However, the decrease of the crystalite size of the bigger particles was more pronounced when the mixing setup "organic-salt" was in use instead of "salt-organic". But it has to be taken into account that different amounts of 5-hexynoic acid were introduced into the reaction zone. Further studies are needed to corroborate a real substantial difference between the mixing setups when using iron nitrate as starting material.

When iron citrate was used, a strong increase in the mean crystallite size from 5 to 33 nm was detected upon the same increase in the reaction temperature from 623 K to 673 K. However, for particles synthesized at a reaction temperature of 623 K, the fit of a bimodal distribution was necessary with the result of a dominant fraction (around 90%) of small crystallites (4.3 nm) and the other fraction containing particles with mean crystallite sizes of 62 nm. At least bimodal distributions were clearly observed for the other mixing setups "organic-salt" and "mixed" at the same reaction temperature with very similar values for the percentage as well the size of the smaller crystallites. However, the size of the bigger particles was significantly and equally smaller indicating that the presence of the carboxylic acids when entering the hot reaction zone influenced the particle formation. Comparing the results of the different mixing setups at a reaction temperature of 673 K mean crystallite sizes of around 30 nm were calculated from the diffraction patterns with an about 10% smaller mean crystallite size for the particles from the "organic-salt" mixing setup and an about 10% larger mean crystallite size for the "salt-organic" setup. The largest value obtained for the mixing setup "salt-organic" at otherwise same reaction conditions points to an influence of the carboxylic acids on the particle formation when entering the hot reaction zone. However, also for iron citrate as starting material further studies are needed to elucidate differences between the mixing setups and the role of the citrate in the particle formation process.

Similar results concerning the order of magnitude of the crystallite sizes and composition were obtained for particles synthesized in the presence of 10-undecynoic acid at a reaction temperature of 673 K. However, comparing the results for the mixing setup, "undecynoic" with the respective results for particles synthesized in the presence of 5-hexynoic acid using the "salt-organic" setup, similar or smaller crystallite sizes were detected. For the particles synthesized using iron nitrate as starting material at an intermediate fraction of maghemite/magnetite (around 75%, in comparision for 5-hexynoic acid: at the highest "organic:salt" ratio about 90% and around 30% for the lower "organic:salt" ratio) the smallest mean crystallite size was determined to be 5.2 nm (in comparision to 8.3 nm and 6.6 nm, respectively). Concerning the mean crystallite size of the hematite particles, a very similar size (about 50 nm) as the one obtained for the higher "organic:salt" ratio was determined in comparison to about 70 nm for the lower ratio. When using iron citrate as starting material, particles synthesized in the presence of 10-undecynoic acid were found to be approximately 20% smaller (mean crystallite size of the maghemite/magnetite particles) than particles synthesized in the presence of 5-hexynoic acid at the same reaction temperature.The smaller crystallite sizes of the particles synthesized in the presence of 10-undecynoic acid in comparison to those synthesized in the presence of 5-hexynoic acid could be an indication for a better stabilization of these particles also under process conditions due to the longer chain length. However, it is also possible that this shift was influenced by the presence of the ethanol and/or the slightly larger flow rates and the introduction of the third cold stream with the consequence of slightly shorter residence times. Further studies are needed to elucidate the influence of the individual factors.

To obtain further insights into the size and morphology of the as-synthesized particles, selected samples were investigated by means of transmission electron microscopy (TEM). Representative images are shown in Figure 7.6. The overview picture of the particles synthesized in Run 4B (left image of Figure 7.6a, mixing setup: "organic-salt", starting material: iron nitrate and 5-hexynoic acid, reaction temperature: 623 K) clearly shows a strongly bimodal distribution congruent with the results obtained for the crystal size analysis by means of PXRD. A large fraction of bigger particles is

**Figure 7.6 *(facing page)*:** TEM images of particles synthesized in the presence of 5-hexynoic or 10-undecynoic acid. In the case of (b) and the left image of (c), as obtained particle dispersions were nebulized onto TEM grids and imaged directly. The particles in the other images underwent rinsing with ethanol to remove excess organic molecules prior to imaging (see also Section 3.2.5 in Chapter 3).

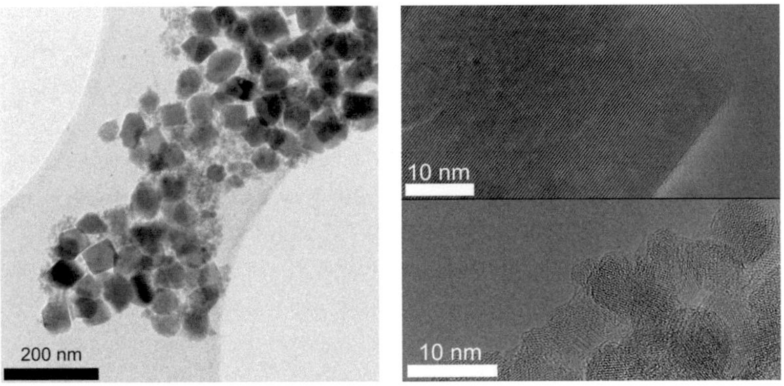


**(a)** Run 4B, "organic-salt", nitrate, 623 K

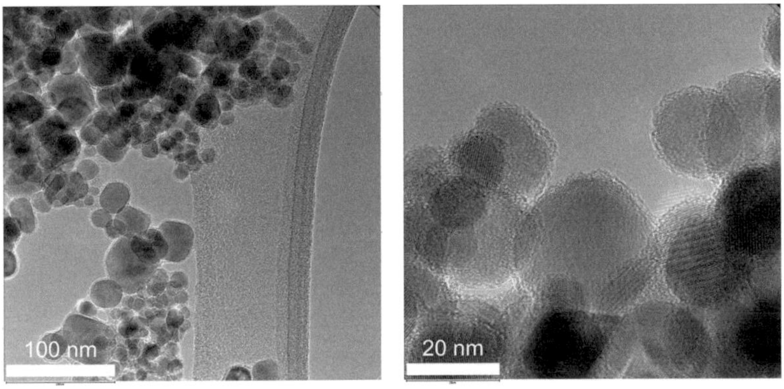


**(b)** Run 5C, "organic-salt", citrate, 673 K

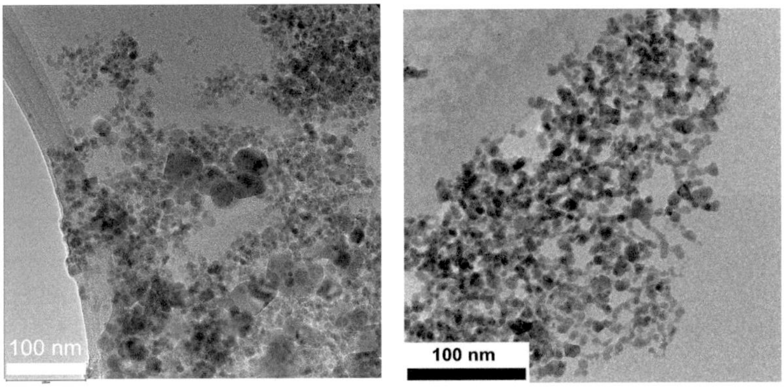


**(c)** Run 7C, "undecynoic", nitrate, 673 K

present. These particles have dimensions predominantely in the size range from 60 to 85 nm (in relative good agreement with the mean crystallite size of 82 nm from PXRD-analysis) and appear mostly as rectangular projections indicating well developed facets. The other particles are distinctively smaller with sizes of one order of magnitude smaller. Images in the high-resolution transmission electron microscopy mode (HRTEM) were taken of a few bigger particles and areas containing smaller particles. For each case, one example is shown next to the overview image. At the top, a representative example of a bigger particle shows the single crystalline structure typical for the larger particles in agreement with the well developed facets. At the bottom, a representative collection of smaller particles is shown. These conglomerates consist of primary particles with fairly similar sizes around 6 nm and roundish shapes. As expected from the PXRD analysis, the pattern of the fast Fourier transform (FFT, data not shown) suggests maghemite or magnetite structure for the smaller particles and hematite for the bigger particles.

Particles synthesized from iron citrate using the same mixing setup – "organic-salt" – at a reaction temperature of 673 K are shown in Figure 7.6b. The overview image on the left side reveals the presence of a broad size range of primary particles from circa 5 nm to 80 nm. Although all sizes in this size range are represented, it looks more like an overlapping bimodal distribution with a fairly broad particle size distribution (one fraction with sizes around 60 nm and the other around 15 nm, based on 200 measured particles). However, compared to the particles sythesized from iron citrate at a reaction temperature of 623 K, the bimodal distribution is not as distinctive. A broadening as well as a converging of the particles sizes could be the reason for the existence of particles over a wide particle size range. Also with respect to the shape, the bigger particles appear not as regular and well faceted as in case of Run 4B. Only a small fraction of the imaged bigger particles showed the distinctive shapes of well grown crystals. However, HRTEM of the crystals (right image) indicates mainly a single crystalline structure of the particles.

A layer is visible at the edge of the particles in the HRTEM images from Run 4B and 5C (right of Figure 7.6a-b). This layer is very likely due to the presence of the organic molecules at the particle surface. However, it is not possible to directly interpret the difference in layer thickness because the sample preparation differs (with vs. without rinsing with ethanol to remove excess organics, for particles from run 4B and 5C, respectively).

At the bottom of the series of TEM images (Figure 7.6c), two overview pictures of particles synthesized in the presence of 10-undecynoic acid using iron nitrate as iron source (Run 7C) are shown. On the left, as-synthesized particles were imaged, whereas on the right the resulting particle size distribution after size selection by centrifugation is shown. Prior to separation (Figure 7.6c, left) a small fraction of well shaped particles of around 40 nm, fairly roundish particles in the size range from 10 to 30 nm and predominantly roundish particles with sizes from 4 to 7 nm were present. Comparing the fraction of the biggest particles with the one obtained in the presence of 5-hexynoic acid (Figure 7.6a) when the same iron source was used, significantly smaller particles were obtained (about half size). After the post treatment (Figure 7.6c, right), the particle size was unambiguously more uniform. Primary particles predominantly in the size range of 4 to 7 nm were revealed, most of them with roundish shape.

Following the removal of larger particles/agglomerates by centrifugation, we found that the remaining particles synthesized in the presence of 10-undecynoic acid formed a stable suspension in ammonia solution (Figure 7.7). This presumably reflected a combination of steric and electrostatic mechanisms due to the formation of a bilayer or multiple layers of 10-undecynoic acid presenting carboxylate ions on the particle surface. TEM images of the particles reveal aggregates/agglomerates of small parti-

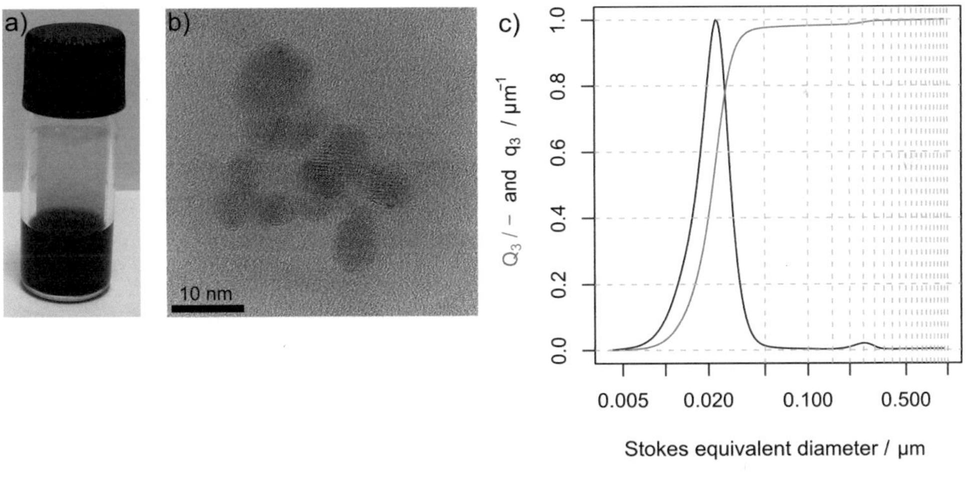


**Figure 7.7:** Characterization of aqueous dispersions of iron oxide nanoparticles functionalized with 10-undecynoic acid (Run 7C). (a) Photograph of dispersion after several months in storage. (b) TEM image of particles from the dispersion. (c) Cumulative and differential particle size distributions based on mass ($Q_3$ and $q_3$) measured with an analytical disc centrifuge.

cles (in the order of 10 nm). The aggregates themselves can be up to 30 nm). Characterization of the particle size distribution (PSD) by analytical disc centrifugation revealed aggregates/agglomerates with a median (based on weight) size of 21.5 nm, with 80% of the particles in the range 13.5–30.5 nm. There was also a small fraction of particles/aggregates/agglomerates with an approximate size of 250 nm.

In addition, the samples were characterized by thermogravimetric analysis in order to obtain a rough estimate of the content of organic modifier in the dispersions and on the particles. For this, the dispersions were introduced without pre-treatment directly into the TGA sample holders. A characteristic TGA scan for the dispersion shown in Figure 7.7 is shown in Figure 7.8. The progression of the TGA scan can be subdivided into three phases:

1. immediate and strong weight loss due to the evaporation of water (more than 90 % of the original mass of the sample up to a temperature of $\approx$ 340 K)
2. (most likely) loss of non chemically bound 10-undecynoic acid up to a temperature of $\approx$ 420 K, after a period of little weight loss
3. small weight loss in the temperature range above 470 K, expected for chemically attached carboxylic acids such as 10-undecynoic acid (e.g. 3,4-dihydroxycinnamic acid on $Fe_3O_4$ nanoclusters starting around 550 K, and for different carboxylic acids on $ZrO_2$ starting around 570 K; see Taguchi et al., 2012; Togashi et al., 2011).

In addition to these losses, changes in the structure of the iron oxide nanoparticles might contribute to the observed weight changes (cf. Schwertmann and Cornell, 2000).

Especially due to the strong and immediate loss of water from the suspensions even before starting the TG analysis, only a rough estimate of the weight fraction of the particles in the suspension can be made—on the order of 1% by mass. The weight loss between $\approx$ 340 K and $\approx$ 420 K ("phase" 2 from above) attributed to unreacted 10-undecynoic acid further supports the hypothesis that a layer of physically adsorbed 10-undecynoic acid is formed on the surface of the particles due to hydrophobic interactions between the hydrocarbon chains. The large amount suggests the formation of multiple layers.

Although the main focus of this work was the demonstration of the *clickability* of the in situ functionalized iron oxide nanoparticles, a very limited number of exploratory measurements of the magnetic properties of the iron oxide nanoparticles

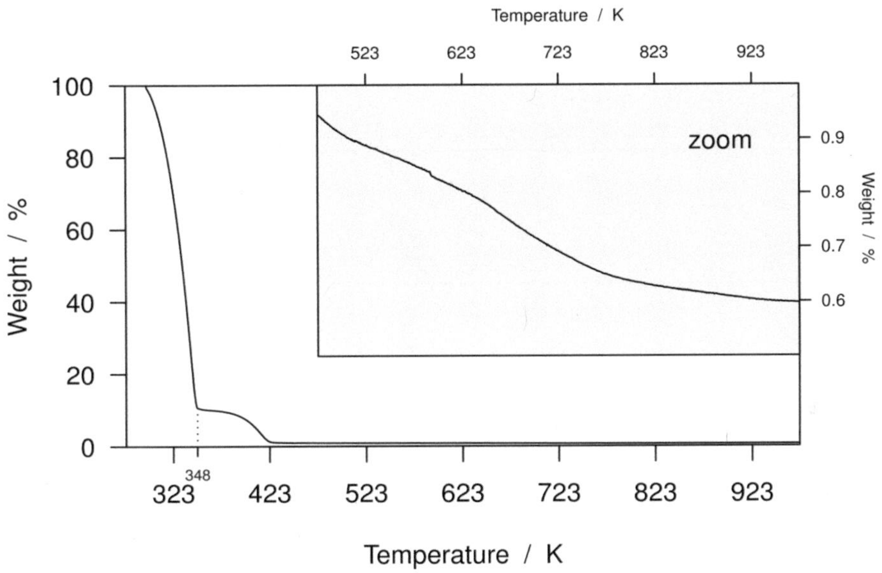


**Figure 7.8:** TGA scan of the dispersion shown in Figure 7.7. The inset is a zoom of the plot region with T > 470 K.

were also carried out using an alternating gradient magnetometer (AGM). Figure 7.9 shows the relationship between applied magnetic field strength and magnetization (magnetization–magnetic field or M–H curve) for particles from Runs 5B and 5C (cf. Table 7.2), measured at room temperature. The saturation magnetization ($M_s$) of the functionalized particles were $\approx$ 30 $\mathrm{A\,m^2\,kg^{-1}}$ and $\approx$ 60 $\mathrm{A\,m^2\,kg^{-1}}$ with a remanent magnetization ($M_r$) of $\approx$ 1.5 $\mathrm{A\,m^2\,kg^{-1}}$ and $\approx$ 10.5 $\mathrm{A\,m^2\,kg^{-1}}$, respectively. In the case of Run 5C, the saturation magnetization of around 60 $\mathrm{A\,m^2\,kg^{-1}}$ is typical for maghemite (60–80 $\mathrm{A\,m^2\,kg^{-1}}$, see Table 2.1 on page 36 and Schwertmann and Cornell, 2000). However, for the particles from Run 5B this value is just about half. This small value may be due to the mostly very small particles from Run 5B (around 5 nm), which means that the fraction of the organic is much higher—a factor that is not considered in the calculation of the characterisitic values. Furthermore, these particles could possess much more defective structures due to their larger surface-to-volume ratio, and the surface itself influences the magnetic properties of the measured particles.

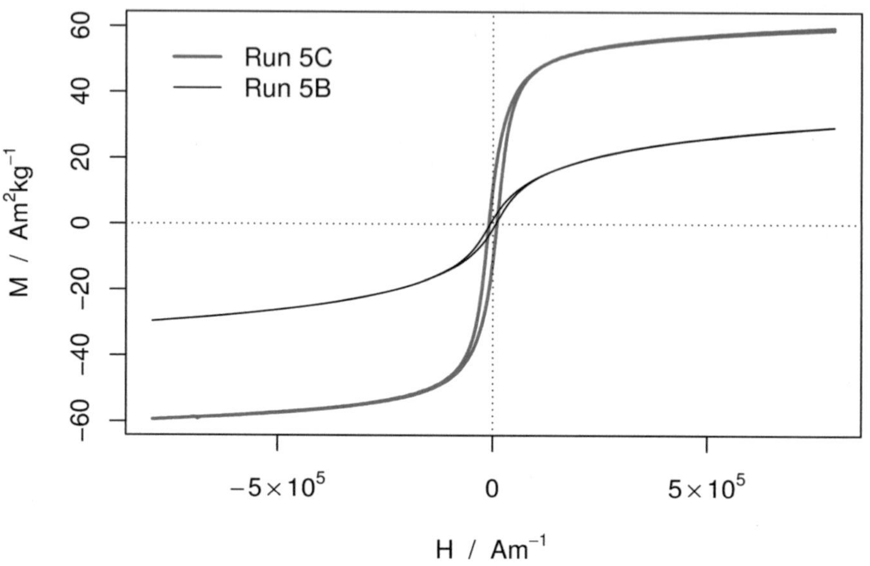


**Figure 7.9:** Magnetization–magnetic field (M–H) curves for particles from Runs 5B and 5C measured at room temperature using an alternating gradient magnetometer (AGM). Three independent samples were measured from Run 5C, whereas particles from Run 5B were measured only once.

Remanence was observed for particles from both runs, with particles from Run 5C exhibiting stronger remanence than particles from Run 5B. This is consistent with the particle size distributions: with a mean crystalite size of around 25 nm and a broad distribution of particle sizes, there is a considerable fraction of particles in Run 5C which is too large to exhibit superparamagnetic behavior. The particles from Run 5B are generally small enough to exhibit superparamagnetic behavior. However, because no size separation was carried out for these particles before performing the measurements of magnetic properties, it is very likely that the small fraction of the larger particles with a mean crystallite size of around 40 nm are responsible for the observed small remanence.

In addition to taking magnetization–magnetic field (M–H) curves for very few samples, a macroscopic test of superparamagnetic behavior was performed with the stable and size selected dispersion from Run 7C shown in Figure 7.7 (cf. Figure 7.6c, right image). Figure 7.10 shows that the particles separate upon being exposed to a strong magnetic field and can be redispersed when the magnet was removed, which macroscopically reveals superparamagnetism behavior.

**Figure 7.10:** Photographs of dispersions (Run 7C; see Figure 7.7), diluted for better visualization; (left) in contact with a strong magnet; (right) after removing the magnet and redispersing.

## 7.4 Characterization of the functionalization

Energy-dispersive X-ray spectroscopy (EDX) combined with imaging in the high angle annular dark field scanning transmission electron microscopy (HAADF-STEM) mode was conducted for particles of Run 4B which consisted of two very different fractions of particles: one portion of mostly well shaped single crystalline particles of sizes around 80 nm and another portion consisting of roundish particles around 6 nm (cf. Figure 7.6a). In Figure 7.11 on the left side, an HAADF-STEM image is shown with two marked areas on which EDX analysis was performed. On the right side, the spectra obtained for the two marked areas are displayed. The upper EDX spectrum was obtained from the section of a large particle (marked "1" on the left side of Figure 7.11) whereas the bottom spectrum was obtained from an area encompassing a part of the surface of a larger particle and several smaller particles (marked

"2"). Both spectra exhibit the characteristic peaks for the expected elements, namely iron and oxygen from the particles, and carbon from the organic molecules bound to their surface, whereby no significant impurities were detected. Comparison of both spectra reveals differences in the relative signal intensity of carbon compared to oxygen and iron: A higher carbon signal was observed for the area marked "2". This observation is consistent with the expectation that the iron oxide nanoparticles are functionalized with the carboxylic acids, yielding a much stronger signal for the smaller particles due to their combined larger surface area. To gain further insights into the functionalization, XPS was carried out for selected samples.

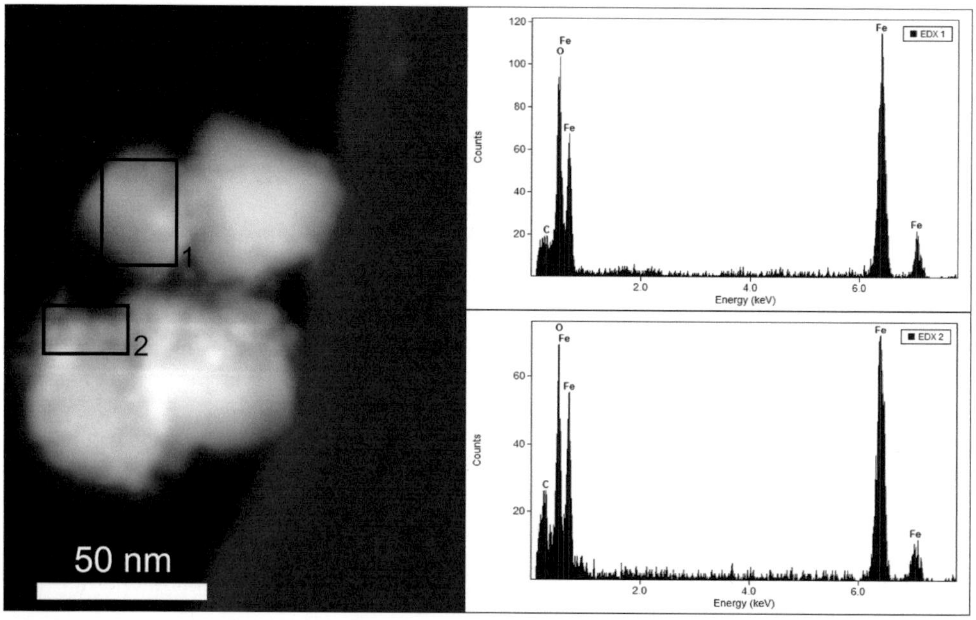


**Figure 7.11:** HAADF-STEM image and EDX spectra of the iron oxide particles from Run 4B, organic-salt, nitrate, 623 K taken at two different spots.

To substantiate the *clickability* of the synthesized particles, we examined particles that had undergone a *click* reaction using X ray photoelectron spectroscopy (XPS). For this, particles synthesized in the presence of 5-hexynoic acid (Run 3B and Run 5B) and 10-undecynoic acid were allowed to react via one of the most used *click* reactions, the Cu(I)-catalyzed azide-alkyne cycloaddition (CuAAC), with the model compound benzyl azide leading to the formation of a 1,2,3-triazole ring. The strategy for the secondary functionalization of the particles is shown in Figure 7.12. For particles synthesized in the presence of 5-hexynoic acid, thoroughly washed and subsequently dried particles were redispersed in DMSO/$H_2O$ in an ultrasound bath, leading only to partial deagglomeration. For the particles synthesized in the presence of 10-undecynoic acid, dispersed particles were used. Here, the step of drying in between was omitted. For all samples, the particles were thoroughly washed to remove any unbound species and dried before carrying out the XPS measurements.

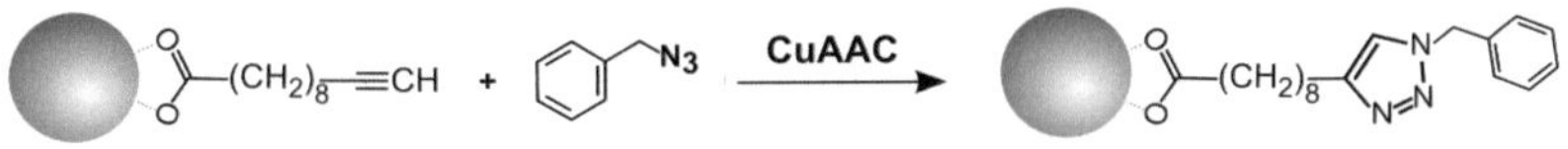


**Figure 7.12:** Strategy for secondary functionalization of particles with *click* chemistry via CuAAC with benzyl azide.

Figure 7.13 shows the N 1s spectra of representative particles synthesized in the presence of 5-hexynoic acid before and after CuAAC. High-quality reference spectra by analyzing pure 5-hexynoic acid, before (measured as the sodium salt) and after the CuAAC reaction, using the same model compound were obtained. As expected, there were no peaks in the N 1s region of the reference samples, but the "clicked" samples (which had undergone CuAAC with benzyl azide) revealed a multiplet comprising at least two peaks from chemically-distinct nitrogen components. On the basis of previous theoretical and experimental studies, this multiplet was always deconvoluted into two peaks and assigned to N=N–N for the peak at around 400.2 eV and to N=N–N for the peak around 402.1 eV, representing the formation of the 1,2,3-triazole ring (Ciampi et al., 2007, 2011; James et al., 2011). The intensity ratio was 1.4:1. XPS spectra were taken after different exposure times to examine the possibility that the reference sample was degraded/transformed during the analysis: No significant changes of the N 1s spectra were observed over the time investigated (refer to Figure F.2 on page 258). Furthermore, similar intensity ratios were also obtained for other reference substances with a high content of the 1,2,3-triazole ring (data not shown). For "clicked" particles (synthesized in the presence of 5-hexynoic acid after CuAAC), fitting of the N 1s multiplet led to two components with similar binding energies and an intensity ratio of $\approx 1.6$.

High resolution scans of the N 1s region for other in situ functionalized particles after CuAAC are shown in Figure 7.14. The characteristic multiplet for the formation of the 1,2,3-triazole ring from 404 eV to 398 eV is observable for all samples investigated. Both the binding energies and intensity ratios agree with published values, which are generally $\leq 2$, with the exception of Run 5B (Ciampi et al., 2007, 2011; James et al., 2011). The observed variations in the intensity ratio, which are also noticeable in the literature, remain to be elucidated. However, the lack of an N 1s component at 404.8 eV for all samples that had undergone CuAAC (representing the electron-deficient nitrogen atom in the azido groups) confirmed that, within the experimental uncertainty, the reaction was completed and no benzylazide was adsorbed nonspecifically (Ciampi et al., 2011; Collman et al., 2006).

The detected degree of secondary functionalization was very different for the samples investigated. This is recognizable by the different signal to noise ratios seen in Figure 7.14. The highest signal to noise ratio was obtained for well dispersed particles of small crystallite sizes (Run 7C, "undecynoic" after centrifugation around 5 nm). Well dispersed particles with larger crystallite sizes (Run 7D, "undecynoic", around 20 nm) showed much lower signal to noise ratios (as expected) due to the much lower surface area. A direct comparison with particles functionalized with hexynoic acid is not possible due to the different treatments these particles underwent prior to the measurements: "hexynoic" particles were dried and resuspended prior to secondary functionalization while "undecynoic" particles were not. Furthermore, differences in dispersability exist due to the different chain length of the primary functionalization. Nevertheless, a secondary functionalization via CuAAC was also demonstrated for particles synthesized in the presence of 5-hexynoic acid (Run 3B, "salt-organic", and Run 5B, "organic-salt").

To further substantiate the finding of *clickable* groups present at the surface of the particles and available for secondary functionalization via CuAAC, we used a *clickable* fluorescent marker ($N_3$-Rhodamine) as secondary ligand and performed the *click* reaction with different in situ functionalized particles (from Runs 3B, 1C and 7D). Concentrated aqueous suspensions of the thoroughly washed (in order to remove excess dye) particles were embedded in mounting medium and examined by laser scanning confocal microscopy. Representative results of this examination are presented in

**Figure 7.13 *(facing page)*:** Investigation of the *clickability* of in situ functionalized particles by XPS using a model compound (benzyl azide). The lines connecting the organic molecules to the surface of the particles are for illustration purposes. The actual binding configuration is not yet known.

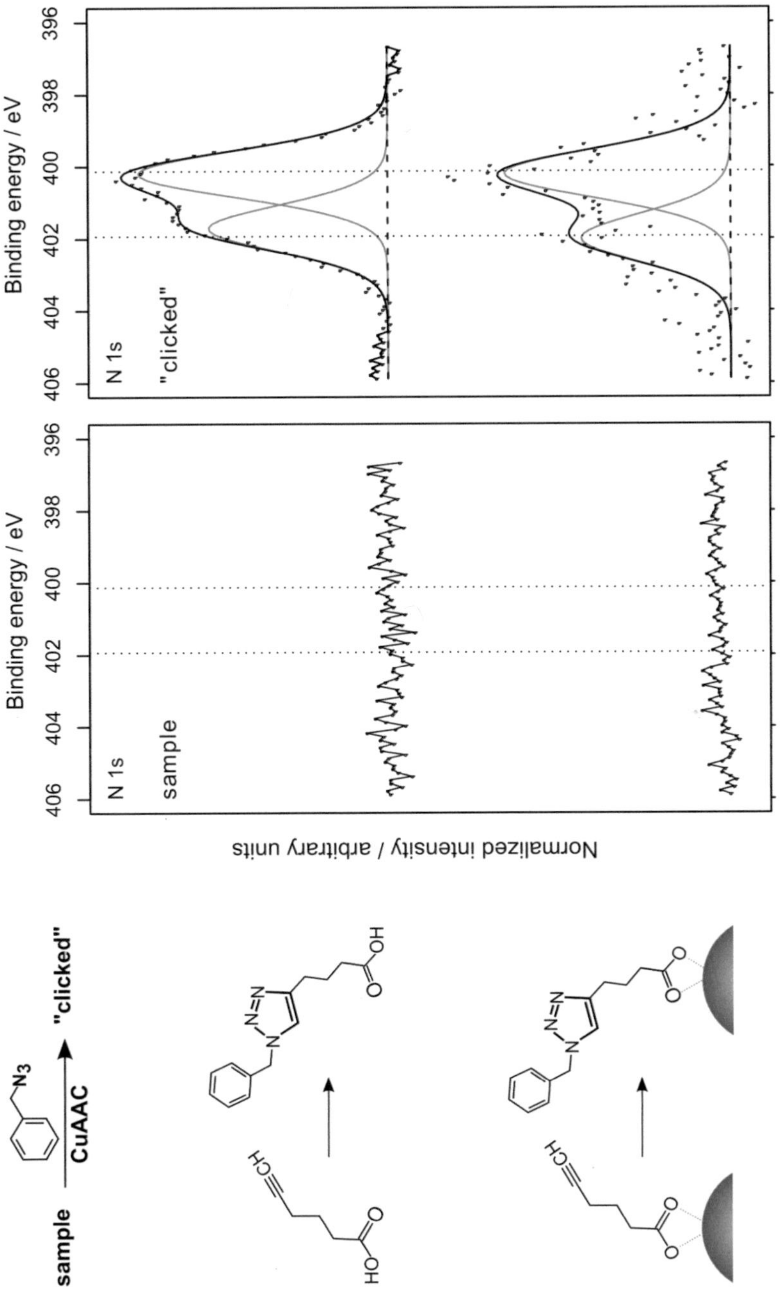

sample
CuAAC
"clicked"
$N_3$
Binding energy / eV
406
404
402
400
398
396
N 1s
sample
"clicked"
Normalized intensity / arbitrary units
HO
OH
N=N
N
O

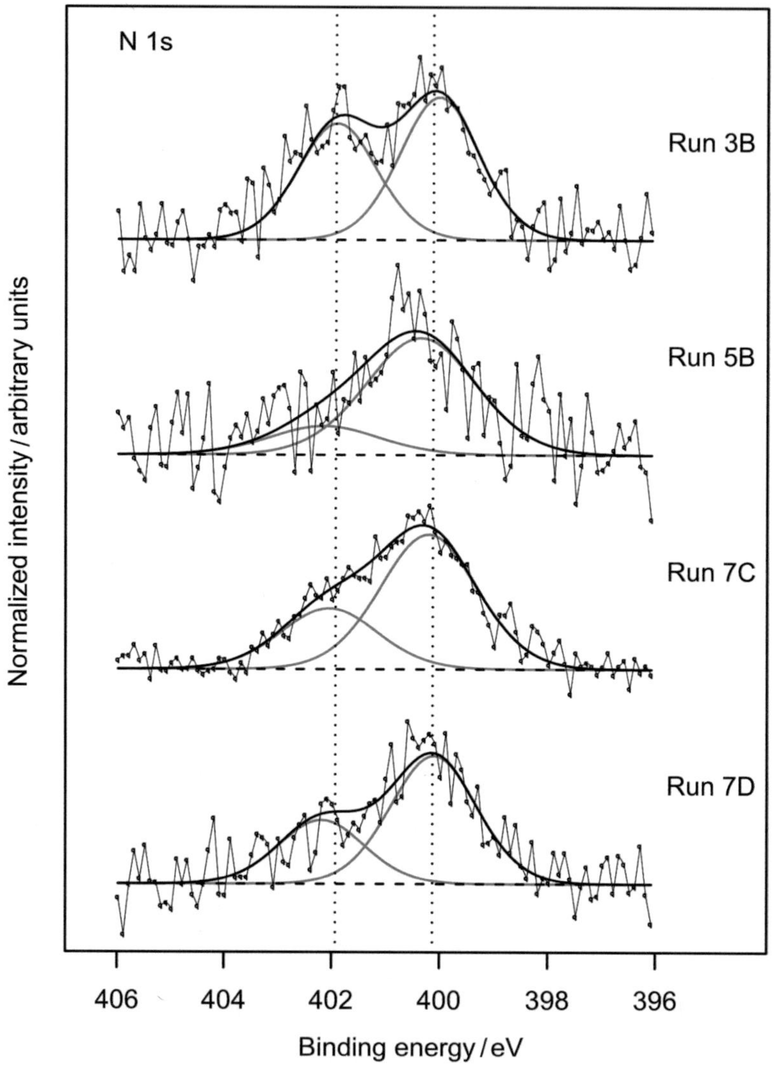


**Figure 7.14:** N 1s – XPS spectra of particles from selected runs which had undergone CuAAC with the model compound benzyl azide.

Figure 7.15 and clearly show that the azide functionalized dye was able to bind to the *clickable* particles (Figure 7.15a, c & d) but not to the control particles (functionalized but *non-clickable*; Figure 7.15b). The fluorescence intensity of the fluorescently-labeled particles was not homogeneous, with dark patches scattered throughout an otherwise strongly-fluorescent background. These patches corresponded to large particle agglomerates remaining after the incomplete deagglomeration of the dried particles (confirmed in the corresponding bright-field image as opaque areas). This result holds true for different chain lengths of the *clickable* anchors as well as for particles decorated with a mixture of *clickable* and *non-clickable* anchors. It is worth noting that no fluorescence was detected in the final washing solution, but very few remaining particle agglomerates could still be found by chance in the last washing solution. Here, the fluorescence of individual larger particle agglomerates was visible against the non-fluorescent background, further confirming the successful functionalization of the particles.

While these experiments provide further evidence of the accesibility of the *clickable* anchors, they cannot be interpreted quantitatively because the particle concentration in the test suspensions is not exactly known, the particles differ in size (see Table 7.2), and the density of available *clickable* anchors (per particle and overall) is not known. Nevertheless, there is a trend in fluorescence intensity in Figure 7.15 that corresponds well with expectations derived from knowledge of the particle sizes. The highest fluorescene intensity was observed for the particles from Run 3B (Figure 7.15a), which are very small ($\approx$ 5 nm) in diameter and were functionalized exclusively with *clickable* anchors. A less intense fluorescence was observed for particles from Run 7D, which were also functionalized exclusively with *clickable* anchors (of longer chain length than Run 3B) but exhibit larger mean particle size ($\approx$ 20 nm) and therefore much smaller specific surface area. Particles from Run 1C showed even lower fluorescence, partly because they present a mixture of *clickable* and *non-clickable* anchors and partly due to the size of the crystallites ($\approx 50\% < 5$ nm and the remainder around 60 nm). Thus, the results provided by the fluorescence images underline the *clickability* of the in situ functionalized particles and point to potential ways of tuning the amounts of bound secondary ligand.

**Figure 7.15 *(facing page)*:** Laser scanning confocal microscopy (LSCM) visualization of the *clickability* of the in situ functionalized particles using an azide-functionalized fluorescent dye (a derivative of Rhodamine B): a) from Run 3B, b) synthesized in the presence of hexanoic acid at 673 K, c) from Run 1C, d) from Run 7D. The lines connecting the organic molecules to the surface of the particles are for illustration purposes. The actual binding configuration is not yet known. The fluorescence image of agglomerates in the washing solution was taken at the same power settings as the other fluorescence images but the contrast was enhanced ex post for clarity.

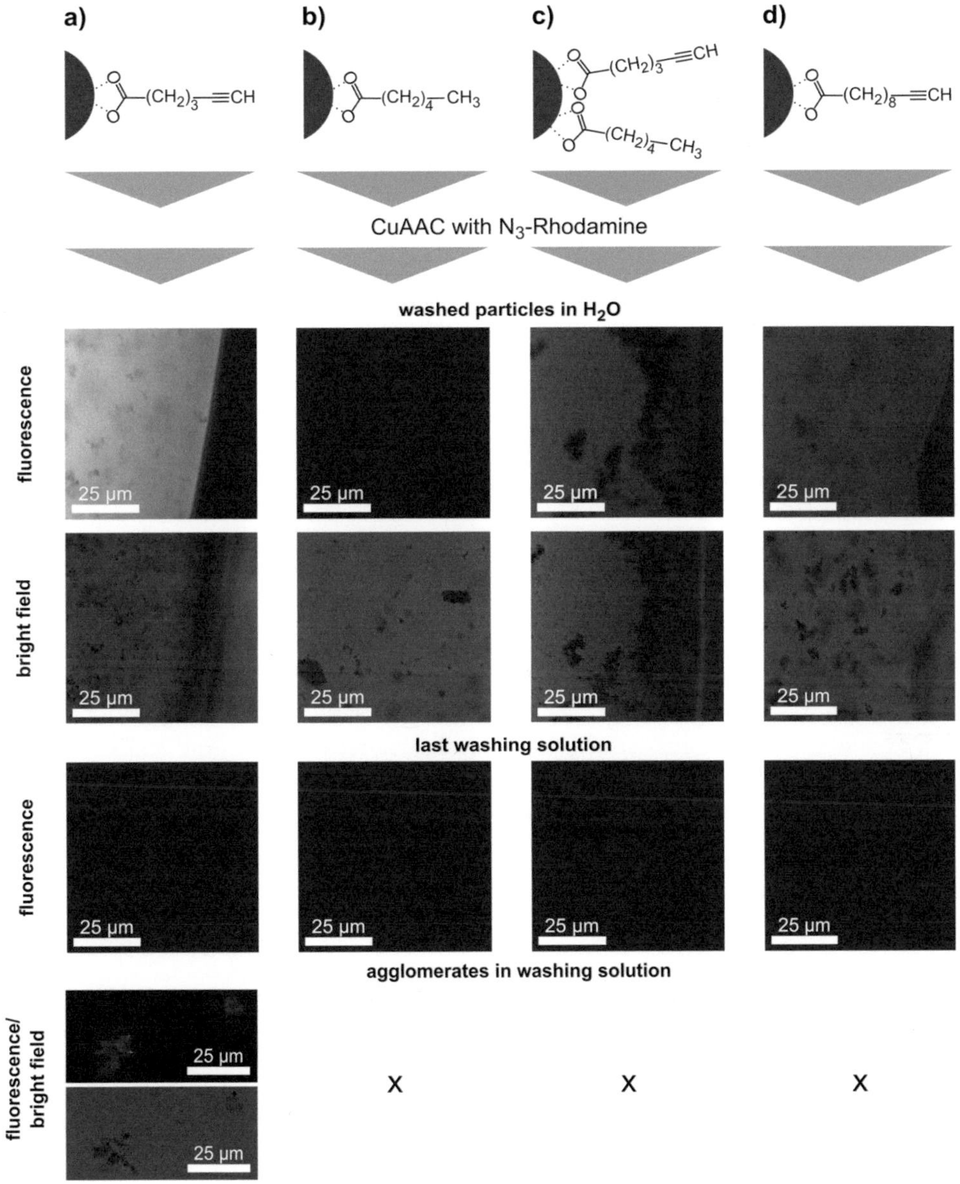

a)
b)
c)
d)
(CH2)3 ≡CH
(CH2)4 CH3
(CH2)8 ≡CH
CuAAC with N3-Rhodamine
washed particles in H2O
fluorescence
bright field
last washing solution
agglomerates in washing solution
fluorescence/ bright field
25 µm
X

## 7.5 Synthesis and characterization of "glyconanoparticles" using galactose with an azide-terminated linker as a model compound

Having shown in principle that it is possible to prepare in situ functionalized, *clickable* iron oxide nanoparticles, the next step was to pursue the preparation of model glyconanoparticles on the basis of the in situ functionalized *clickable* iron oxide nanoparticles. For this, a monosaccharide (galactose) with an azide linker was chosen (see Figure 7.16). The synthesis and purification of the galactose with an azide-terminated linker was developed and carried out by Stefan Jennewein and his group at Fraunhofer IME and kindly provided for these investigations.

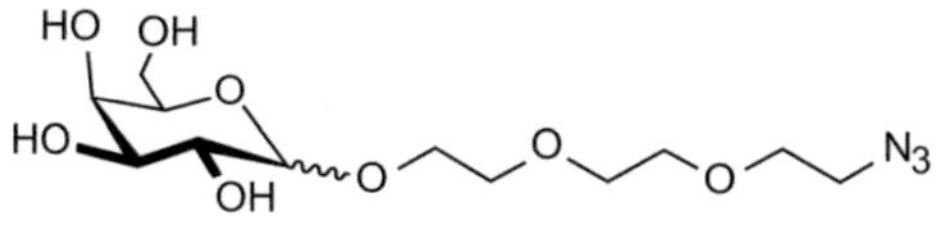


**Figure 7.16:** Structure of galactose with azide-terminated linker.

Particles synthesized in the presence of 10-undecynoic acid (Run 7C) were allowed to react via the CuAAC with the galactose possesing the azide-terminated linker. The procedure for the synthesis and purification of the model glyconanoparticles was very similar to the procedure described in the previous section for the primary functionalized particles that were allowed to react with the model compound benzyl azide. These as-prepared particles were examined by means of XPS to probe the (covalent) attachment of the galactose to the surface of the particles presenting the clickable anchors. Figure 7.17 shows representative N 1s spectra of the galactose derivative before the attachement (a) and after the CuAAC with the primary functionalized particles (b).

The high resolution scans of the N 1s region for the galactose with the azide-terminated linker revealed the two characteristic distinct peaks for organic azides, one centered at around 404.8 eV and the other at around 401.2 eV. The smaller peak represents the electron-deficient nitrogen atom in the azido groups. However, as the small peak was also always observable in the lower binding energy region, as seen in Figure 7.16a. This peak is also present to a greater or lesser extent upon close examination of published N 1s spectra of organic azides (e.g., Collman et al., 2006; Wollman

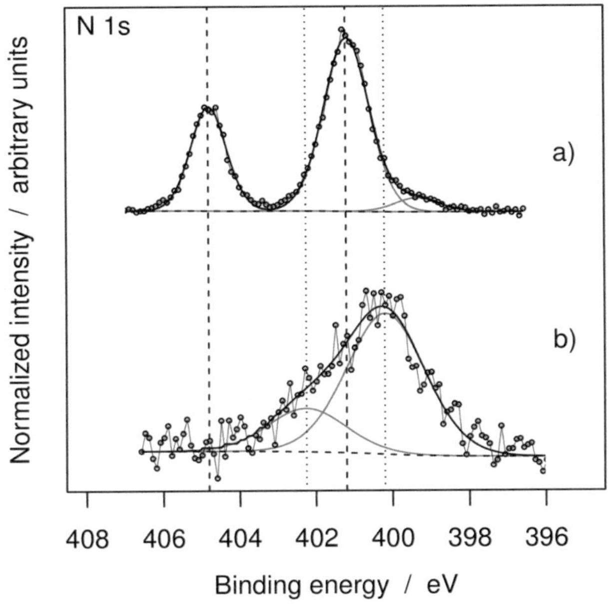


**Figure 7.17:** N 1s – XPS spectra of (a) galactose with azide-terminated linker (cf. Figure 7.16 for structure) and (b) of particles from Run 7C which had undergone CuAAC with this compound.

et al., 1994), pointing to potential degradation/transformation during XPS-analysis. Therefore, the galactose with the azide-terminated linker was investigated via XPS after different exposure times. An overlay of the scans is shown in Figure F.3 on page 259. The fraction of the peak area from the "additional"peak at lower binding energy is constantly changing indicating that some kind of transformation is occuring. Other organic azide compounds have also been investigated with the same procedure showing a continuous transformation of the XPS-pattern with the three peaks in favor of the peak at lower binding energy upon increasing exposure time. That this is not observed for the galactose with the azide-terminated linker is most likely related to the fact that this compound—in contrast to the other organic azides investigated*—was not measured as a solid but as a highly viscous liquid, which means that new material can enter the measurement spot. Combining these results, it may be concluded that the organic azides are X-ray sensitive, leading to artefacts related

*Personal communication from Michael Bruns and Leonie Barner, KIT.

to exposure time when analyzed via XPS. Furthermore, the peaks and the transformation pattern observed are characteristic for organic azides.

The high resolution scan of the N 1s region of primary functionalized particles that were allowed to react with the galactose with the azide-terminated linker via CuAAC revealed the characteristic multiplet of the 1,2,3-triazole ring from 404 eV to 398 eV (see Figure 7.16b), indicating the successful *click* reaction between surface bound alkynes and the terminal azide group of the galactose compound (cf. Section 7.4).

## 7.6 Interim summary

Iron oxide nanoparticle dispersions of varying properties (e.g. color, crystal structure, particle size distribution) were obtained in a continuous hydrothermal pilot plant by varying the mixing setup, reaction temperature and the starting material (iron salt) in the presence of different carboxylic acids with the aim of producing in situ functionalized nanoparticles suitable for secondary functionalization via *click* chemistry. The resulting nanoparticles could be at least partially transferred to toluene—an indication of the association of the organic modifier with the surface of the iron oxide nanoparticles—depending on the synthesis conditions. The degree of transfer from the aqueous to the organic phase correlated well with the length of the organic chains and the identity of the terminal functional group.

The stability of the *clickable* ligands in the harsh hydrothermal environment was also tested. These experiments revealed that the alkyne functional group is sufficiently stable under the investigated conditions. When present at the surface of the iron oxide nanoparticles, the alkynes were shown to be available for CuAAC reactions. The versatility of the continuous hydrothermal process in terms of the metal oxides that can be synthesized combined with the versatility in functionalization introduced by using *click* chemistry could enable a wide variety of applications in diverse fields. Nanoparticles presenting galactose were prepared as a model glyconanoparticle—a successful demonstration of the process using a potentially relevant surface functionalization.

# Chapter 8

# Summary and future directions

Near- and supercritical water is an attractive medium for the synthesis of metal oxide nanoparticles. The continuous hydrothermal synthesis process is based on the mixing of a cold aqueous metal salt solution with hot compressed water which has been investigated for more than 20 years (Adschiri et al., 1992). This process exploits the strong variations in the properties of water — in particular its permittivity — around the critical point ($p$ = 22.1 MPa, $T$ = 647.1 K). In this regime, water has very unusual and tunable properties. Macroscopically it behaves like non-polar solvents but the single molecules are still polar (Kruse and Dinjus, 2007). These special and strongly variable properties of water make near-critical and supercritical water also a suitable medium for the synthesis of in situ functionalized metal oxide nanoparticles, providing further control of particle formation regarding size, morphology and surface chemistry as compared to the supercritical hydrothermal synthesis of non-functionalized particles (e.g. Byrappa et al., 2008; Lu et al., 2012; Mousavand et al., 2006; Taguchi et al., 2012; Ziegler et al., 2001).

Both the hydrothermal synthesis of functionalized and non-functionalized nanoparticle dispersions were pursued in this thesis. The goals of this work were:

1. To improve the understanding of the the relationship between different process variables and the resulting nanoparticle dispersions.
2. To extend the usability of this process by introducing *clickable* organic modifiers, thereby contributing to the development of a convenient and versatile process for the synthesis of metal oxide nanoparticles with universal and accessible anchors on their surface.
3. To show a simple route for the preparation of glyconanoparticles.

The basis for approaching these goals was the design, construction, and optimization of a pilot plant for the continuous hydrothermal synthesis of non-functionalized and functionalized metal oxide nanoparticles under near-critial and supercritical conditions. Consequently, a pilot plant was built operating at $T_{max}$ = 773 K and $p_{max}$ = 38 MPa with a functioning process measurement, control, and data logging system accomodating the special safety issues related to operation under high temperature and pressure. In the detailed design of the high pressure pilot plant, special emphasis was given to following aspects:

- The realization of high total mass flow rates ($\dot{m}_{max}$ = 12 kg h$^{-1}$) compared to the pilot plants used by other reaserch groups. This enabled very short residence times and the investigation of the influence of variations in the total mass flow rate on particle formation over a very wide range of mean flow velocities.
- The easy adjustment of the mixing and reactor geometry. This allowed for the introduction of (additional) streams at different positions and geometries in the reaction zone.
- The operation of the pilot plant with low pulsation (because of the sensitivity of the reaction conditions to even small changes in pressure).

All three requirements were fulfilled with the pilot plant by a combination of design and the choice of high pressure pumps with essentially no pulsation. However, a fully automated pressure control system is still missing and should be installed. Moreover, the seals in the (high pressure) cold metal salt line often failed, forcing an interruption of the operation of the pilot plant. Failure in the seals for the organic stream was much less frequent. Therefore, there is a need to identify sealing materials that will enable a long term operation of a continuous hydrothermal synthesis plant. Finally, the pilot plant was not designed to minimize energy consumption, an aspect that should be considered in future work to reduce the energy needed for the realization of the upscaled process.

The synthesis of cerium oxide was chosen as a model system to gain details on the performance of the newly built pilot plant. These preliminary experiments were meant (a) to test the reproducibility and reliability of the process with replicate runs at the same nominal conditions and (b) to test the reliability and stability of the process at different cerium content. For this, the high pressure pilot plant was first tested by synthesizing cerium oxide from aqueous cerium nitrate solutions. As a result of these experiments, electrostatically stabilized cerium oxide nanoparticle dispersions were obtained at conversions $>$ 99.8% with residence times in the reactor $<$ 1 s. The cerium oxide nanoparticles had mean crystallite sizes in the range of 30-45 nm by PXRD. By means of differential centrifugal sedimentation analysis (DCS), a technique both powerful and convenient for obtaining particle/agglomerate size distributions, these crystallites were shown to be mostly singly dispersed at low cerium content and to form agglomerates and/or aggregates with increasing cerium content. However, the size of the cerium oxide crystallites did not show a dependence on the cerium content of the suspensions (varied over three orders of magnitude) at the conditions investigated.

The relatively poor reproducibility concerning the mean crystallite size observed in the replicate runs and sporadical clogging of the mixing unit signaled the need for optimizing the mixing geometry of the pilot plant. In later experiments on the synthesis of iron oxide nanoparticle dispersions, a mixer (T-unit) with invariant inner diameter was introduced to replace the standard high pressure T-unit, which have abrupt changes in their inner diameter and were used for the cerium oxide experiments. The result was less frequent clogging of the pilot plant and, in combination with other measures related to the relative flow rates of the process streams, helped to ensure a reliable long term operation of the pilot plant. This approach appears promising for the future improvement of the synthesis of electrostatically stabilized cerium oxide dispersions in a continuous hydrothermal process.

Following the experiments with cerium oxide, electrostatically stabilized iron oxide nanoparticle dispersions were prepared from iron(II) acetate, iron(III) nitrate, and mixtures of the two, with or without the addition of the respective acids ($CH_3COOH$ and $HNO_3$ respectively). The dispersions were characterized by DCS, electrophoretic light scattering, TEM, PXRD, and by Raman spectroscopy using a modified setup that allows measurements directly in concentrated dispersions. Using these techniques, it was found that dispersions synthesized from pure aqueous iron(III) nitrate solutions were bimodal regarding both primary particles as well as agglomerates. However, upon increasing the total mass flow rate and mixing ratio between the aqueous metal salt stream and supercritical water stream, the fraction of the larger particles and agglomerates (around 30 nm and 65 nm, respectively) was decreased in favor of the smaller particles and agglomerates (around 7 nm and 20 nm, respectively) up to a point were essentially no larger particles were present. Moreover, a mixture of $\alpha$-$Fe_2O_3$ and $\gamma$-$Fe_2O_3$ and/or ferrihydrite were detectable when investigating the dispersions by means of Raman spectroscopy. It is important to note that the structure and size of the synthesized particles can change upon drying, as shown by combining analyses of the dispersions with analyses of the particles as dry powders. In consequence, potentially misleading results are obtained when the particles are measured only as dried powders. However, further Raman spectroscopic investigations of iron oxides in dispersions are needed to discriminate and obtain more information regarding structure and size of the particles in dispersed form. Such insights are of interest beyond the field of hydrothermal synthesis due to the overall importance of iron oxides/oxyhydroxides.

Significantly narrower particle/agglomerate size distributions with primary particles $< 10$ nm most likely of uniform crystal structure could be obtained when acidified mixtures of iron(II) acetate and iron(III) nitrate were used as a starting material instead of pure iron(III) nitrate under the same flow conditions. Furthermore, the experiments on systematically changing the composition of the starting material revealed the potential of varying the primary particle and agglomerate size distribution of the obtained dispersions continuously. Although, further work is needed to discriminate between the different effects that can lead to the shift of the particle size distributions as observed with DCS, it is clear that the educt composition has a strong influence on the product properties. Thus, it seems reasonable to screen for favorable compositions of starting material as an additional variable in optimizing the hydrothermal synthesis of metal oxide nanoparticles. This argument is strengthened by the finding that the composition of the educt solutions also has an influence on the sensitivity of the product properties to fluctuations in the process/flow conditions—an important aspect for robust up-scaling. Furthermore, the use of a highly sensitive material system such as iron oxides/oxyhydroxides together with methods for analyzing the particles/particle agglomerates directly in the dispersions provides a good basis e.g. for studies of mixing in continuous hydrothermal synthesis, allowing both the comparison of different mixing geometries and the optimization of the process regarding reproducibility. Future studies could combine the methods described here with ongoing research on optimizing the mixing process by specially designed high pressure micro-mixing units in order to further narrow the particle size distributions and increase the concentration of metal oxides in stable product dispersions. In addition, an extension of the work shown here to include a study varying the mass flow rate and residence time independently is desirable to improve the understanding of the influence of process conditions on the product properties. Such a study should be accompanied by advances in modelling of particle formation and growth considering the underlying complex phase behavior and the kinetics of the relevant reactions in the unusual environment provided by near-critical and supercritical water.

Following the preparation of electrostatically stabilized iron oxide nanoparticle dipserions, iron oxide nanoparticle dispersions of varying properties (e.g. color, crystal structure, particle size distribution) were prepared in the continuous hydrothermal pilot plant by varying the mixing setup, reaction temperature and the starting material (iron salt) in the presence of different carboxylic acids with the aim of producing in situ functionalized nanoparticles suitable for secondary functionalization via *click* chemistry. The stability of the *clickable* ligands (alkynoic acids) in the harsh hydrothermal environment was also tested. These experiments revealed that the alkyne functional group is sufficiently stable under the investigated conditions, but detailed studies are still needed to elucidate the decomposition pathways and possibilities to further stabilize the alkynoic acids used for in situ functionalization.

The particles synthesized in the presence of carboxylic acids could be at least partially transfered to toluene, an indication of the association of the organic modifier with the surface of the iron oxide nanoparticles. The degree of transfer from the aqueous to the organic phase correlated well with the length of the organic chains and the identity of the terminal functional group. To confirm the presence of *clickable* functional groups for particles synthesized in the presence of alkynoic acids, copper(I)–catalyzed azide-alkyne cycloaddition (CuAAC) was carried out with the particles and an azide-functionalized fluorescent dye (a derivative of Rhodamine B). Particles that had not undergone CuAAC with the fluorescence dye and particles functionalized with hexanoic acid (without alkyne functional groups) were used as negative controls. Concentrated aqueous suspensions of the thoroughly washed particles were examined by laser scanning confocal microscopy, revealing a strong fluorescent signal for the *clickable* particles but no signal for the control samples. Intermediate fluorescence intensities were obtained from particles synthesized in the presence of a mixture of *clickable* and *non-clickable* ligands, as well as from larger particles (lower specific surface area). To further substantiate the *clickability* of the synthesized particles, the formation of the 1,2,3-triazole ring following CuAAC of the in situ functionalized particles with a model compound (benzyl azide) was investigated using X-ray photoelectron spectroscopy. This examination revealed a characteristic multiplet comprising at least two peaks from chemically-distinct nitrogen components (N=N–N and N=N–N), indicating the formation of the 1,2,3-triazole ring. This compared well with positive control samples and no comparable signal could be detected from negative control experiments.

Having shown in principle that it is possible to prepare in situ functionalized, *clickable* iron oxide nanoparticles, the next step was to pursue the preparation of glyconanoparticles on the basis of the primary functionalized iron oxide nanoparticles.

Glyconanoparticles are a class of hybrid nanoparticles that are of increasing importance due to the enormous but for a long time underestimated potential of oligosaccharides. These develop various biological functions when clustered at a surface. Examples for potential applications of glyconanoparticles are anti-cancer vaccines Ojeda et al. (2007), anti-viral agents Martínez-Ávila et al. (2009), and in vivo diagnosis (van Kasteren et al., 2009). Within this work, the preparation of glyconanoparticles via *click* reactions between the alkyne-terminated iron oxide nanoparticles and monosaccharides (galactose) bearing an azide-terminated linker was demonstrated.

Placing the results on the synthesis of in situ functionalized *clickable* iron oxide nanoparticles in the context of the versatility of the continuous hydrothermal process in terms of the metal oxides that can be synthesized and the types of molecules accessible via *click* chemistry, this approach could enable the production of a very wide range of core-shell nanoparticles of known, envisioned and yet-to-be discovered properties. Hybrid organic–inorganic nanostructured materials have potential application in various technologies related to water and energy supply and efficiency, environmental protection, health, micro-electronics, optics and transportation Beija et al. (2011); Sanchez et al. (2011); Yin and Alivisatos (2005). Selected potential forms of use are shown in Figure 8.1, where the contributions of this thesis to the preparation of hybrid core-shell nanoparticles are summarized.

There remain, however, a number of open questions and opportunities for extension of the results in this thesis: It is still not clear by what mechanism the carboxylic acids are bound to the surface of the iron oxide nanoparticles, nor could the amount of bound carboxylic acids and available *clickable* groups be precisely quantified in this work. Moreover, the factors affecting the ligand density on the surface of the particles, such as pH, identity of the counterion and the use of functional groups other than carboxylic acids (e.g. hydroxyl groups), still need to be explored. It is clear that the potential for control of particle formation and growth through in situ functionalization and optimized mixing has not yet been exhausted. Furthermore, the transfer of this approach to material systems other than iron oxides would extend the usefulness of the process demonstrated here. A further avenue with large potential is the in situ functionalization of the particles with more than one type of *clickable* ligand: This would enable multiple secondary functionalization and by that a wider range of applications and a finer control of the surface chemistry of the functionalized particles because of the orthogonality of *click* reactions.

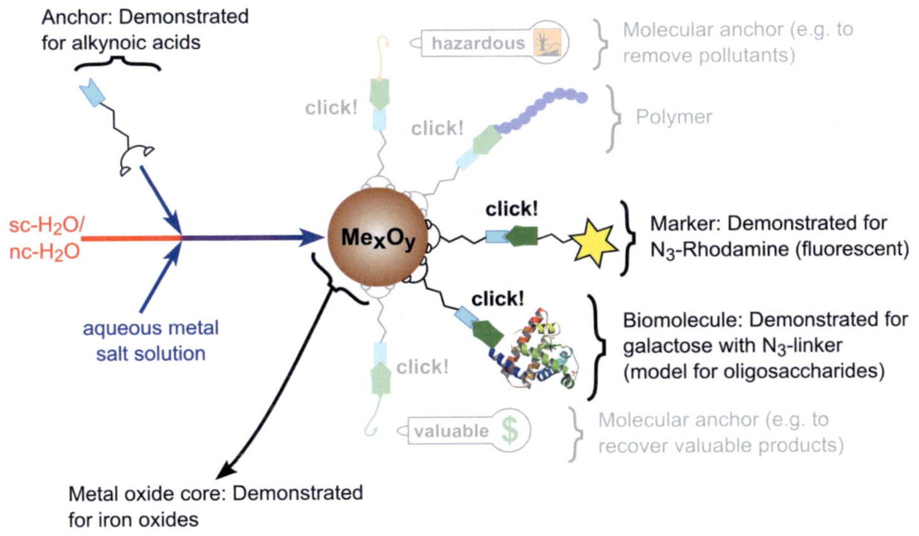


**Figure 8.1:** Schematic representation of the process for the synthesis of metal oxide nanoparticles presenting *clickable* anchors and selected potential forms of use specifying and highlighting the facets covered in this thesis.

# Bibliography


Adschiri, T., Arai, K., 2002. Hydrothermal synthesis of metal oxide nanoparticles under supercritical conditions. New York, pp. 311–326.

Adschiri, T., Hakuta, Y., Arai, K., 2000. Hydrothermal synthesis of metal oxide fine particles at supercritical conditions. Industrial & Engineering Chemistry Research 39 (12), 4901–4907.

Adschiri, T., Hakuta, Y., Sue, K., Arai, K., 2001. Hydrothermal synthesis of metal oxide nanoparticles at supercritical conditions. Journal of Nanoparticle Research 3 (2-3), 227–235.

Adschiri, T., Kanazawa, K., Arai, K., 1992. Rapid and continuous hydrothermal crystallization of metal oxide particles in supercritical water. Journal of the American Ceramic Society 75 (4), 1019–22.

Adschiri, T., Zhang, J., Mousavand, T., Hayashi, N., Ohara, S., Umetsu, M., Naka, T., 2006. Supercritical hydrothermal synthesis of organic-inorganic hybrid nanocrystals and its perfect dispersion and selfassembly. In: 3rd International Meeting on High Pressure Chemical Engineering (HPCE-3). VDI-GVC-Fachausschuss Hochdruckverfahrenstechnik, Erlangen, Germany.

Adschiri, T., Zhang, J., Tahereh, M., Ohara, S., Umetsu, M., Naka, T., Takami, S., 2005. Supercritical hydrothermal synthesis of organic-inorganic hybrid nanoparticles and its perfect dispersion in solvents. In: 10th European Meeting on Supercritical Fluids. Strasbourg/Colmar, France.

Adzima, B. J., Bowman, C. N., 2012. The emerging role of click reactions in chemical and biological engineering. AIChE Journal 58 (10), 2952–2965.

Aimable, A., Muhr, H., Gentric, C., Bernard, F., Le Cras, F., Aymes, D., 2009. Continuous hydrothermal synthesis of inorganic nanopowders in supercritical water: Towards a better control of the process. Powder Technology 190 (1-2), 99–106.

Aizawa, T., Masuda, Y., Minami, K., Kanakubo, M., Nanjo, H., Smith, R. L., 2007. Direct observation of channel-tee mixing of high-temperature and high-pressure water. The Journal of Supercritical Fluids 43 (2), 222–227.

Arita, T., Ueda, Y., Minami, K., Naka, T., Adschiri, T., 2010. Dispersion of fatty acid surface modified ceria nanocrystals in various organic solvents. Industrial & Engineering Chemistry Research 49 (4), 1947–1952.

Ashcroft, N., Mermin, N., 1976. Solid state physics. Saunders College Publishing.

Aymonier, C., Loppinet-Serani, A., Reverón, H., Garrabos, Y., Cansell, F., 2006. Review of supercritical fluids in inorganic materials science. The Journal of Supercritical Fluids 38 (2), 242–251.

Balbuena, P. B., Johnston, K. P., Rossky, P. J., 1994. Molecular simulation of a chemical reaction in supercritical water. Journal of the American Ceramic Society 116 (6), 2689–2690.

Balmer, J. A., Le Cunff, E. C., Armes, S. P., Murray, M. W., Murray, K. A., Williams, N. S. J., 2010. When does silica exchange occur between vinyl polymer-silica nanocomposite particles and sterically stabilized latexes? Langmuir 26 (16), 13662–13671.

Beattie, I. R., Gilson, T. R., 1970. The single-crystal raman spectra of nearly opaque materials. Iron(III oxide and chromium(III) oxide. Journal of the Chemical Society A: Inorganic, Physical, Theoretical, 980–986.

Beija, M., Marty, J.-D., Destarac, M., 2011. RAFT/MADIX polymers for the preparation of polymer/inorganic nanohybrids. Progress in Polymer Science 36 (7), 845–886.

Bell, N. C., Minelli, C., Tompkins, J., Stevens, M. M., Shard, A. G., 2012. Emerging techniques for submicrometer particle sizing applied to Stöber silica. Langmuir 28 (29), 10860–10872.

Belsky, A. J., Maiella, P. G., Brill, T. B., 1999. Spectroscopy of hydrothermal reactions 13. Kinetics and mechanisms of decarboxylation of acetic acid derivatives at 100-260°C under 275 bar. The Journal of Physical Chemistry A 103 (21), 4253–4260.

Bersani, D., Lottici, P. P., Montenero, A., 1999. Micro-raman investigation of iron oxide films and powders produced by sol-gel syntheses. Journal of Raman Spectroscopy 30 (5), 355–360.

Bicker, M., Endres, S., Ott, L., Vogel, H., 2005. Catalytical conversion of carbohydrates in subcritical water: A new chemical process for lactic acid production. Journal of Molecular Catalysis A: Chemical 239 (1-2), 151–157.

Binder, W. H., Sachsenhofer, R., 2008. "Click" chemistry in polymer and material science: An update. Macromolecular Rapid Communications 29 (12-13), 952–981.

Blood, P. J., Denyer, J. P., Azzopardi, B. J., Poliakoff, M., Lester, E., 2004. A versatile flow visualisation technique for quantifying mixing in a binary system: Application to continuous supercritical water hydrothermal synthesis (SWHS). Chemical Engineering Science 59 (14), 2853–2861.

Bondarenko, G. V., Gorbaty, Y. E., Okhulkov, A. V., Kalinichev, A. G., 2006. Structure and hydrogen bonding in liquid and supercritical aqueous NaCl solutions at a pressure of 1000 bar and temperatures up to 500°C: A comprehensive experimental and computational study. The Journal of Physical Chemistry A 110 (11), 4042–4052.

Bremholm, M., Becker-Christensen, J., Iversen, B. B., 2009a. High-pressure, high-temperature formation of phase-pure monoclinic zirconia nanocrystals studied by time-resolved in situ synchrotron X-ray diffraction. Advanced Materials 21 (35), 3572–3575.

Bremholm, M., Felicissimo, M., Iversen, B. B., 2009b. Time-resolved in situ synchrotron X-ray study and large-scale production of magnetite nanoparticles in supercritical water. Angewandte Chemie International Edition 48 (26), 4788–4791.

Bremholm, M., Jensen, H., Iversen, S. B., Iversen, B. B., 2008. Reactor design for in situ X-ray scattering studies of nanoparticle formation in supercritical water syntheses. The Journal of Supercritical Fluids 44 (3), 385–390.

Brundle, C., Evans, C., Wilson, S., 1992. Encyclopedia of Materials Characterization. Manning Publications.

Brunner, G., 2009. Near critical and supercritical water. Part I. Hydrolytic and hydrothermal processes. The Journal of Supercritical Fluids 47 (3), 373–381.

Butenhoff, T. J., Goemans, M. G. E., Buelow, S. J., 1996. Mass diffusion coefficients and thermal diffusivity in concentrated hydrothermal $NaNO_3$ solutions. The Journal of Physical Chemistry 100 (14), 5982–5992.

Byrappa, K., Adschiri, T., 2007. Hydrothermal technology for nanotechnology. Progress in Crystal Growth and Characterization of Materials 53 (2), 117–166.

Byrappa, K., Ohara, S., Adschiri, T., 2008. Nanoparticles synthesis using supercritical fluid technology - towards biomedical applications. Advanced Drug Delivery Reviews 60 (3), 299–327.

Cabanas, A., Poliakoff, M., 2001. The continuous hydrothermal synthesis of nanoparticulate ferrites in near-critical and supercritical water. Journal of Materials Chemistry 11, 1408–1416.

Callister, J. W. D., 1997. Materials Science and Engineering – An Introduction, 4th Edition. John Wiley & Sons, Inc., New York.

Cardona, M., Ley, L., 1978-1979. Photoemission in Solids. Vol. 1 and 2. Springer.

Chen, B., Wu, W., Wang, X., 2011. Magnetic iron oxide nanoparticles for tumor-targeted therapy. Current Cancer Drug Targets 11, 184–189.

Chernyshova, I. V., Hochella Jr., M. F., Madden, A. S., 2007. Size-dependent structural transformations of hermatite nanoparticles. 1. Phase transition. Physical Chemistry Chemical Physics 9 (14), 1736–1750.

Chourpa, I., Douziech-Eyrolles, L., Ngaboni-Okassa, L., Fouquenet, J.-F., Cohen-Jonathan, S., Souce, M., Marchais, H., Dubois, P., 2005. Molecular composition of iron oxide nanoparticles, precursors for magnetic drug targeting, as characterized by confocal raman microspectroscopy. The Analyst 130 (10), 1395–1403.

Ciampi, S., Böcking, T., Kilian, K. A., James, M., Harper, J. B., Gooding, J. J., 2007. Functionalization of acetylene-terminated monolayers on Si(100) surfaces: A click chemistry approach. Langmuir 23 (18), 9320–9329.

Ciampi, S., James, M., Michaels, P., Gooding, J. J., 2011. Tandem "click" reactions at acetylene-terminated Si(100) monolayers. Langmuir 27 (11), 6940–6949.

Cipolla, L., Peri, F., Airoldi, C., 2008. Glycoconjugates in cancer therapy. Anti-Cancer Agents in Medicinal Chemistry (Formerly Current Medicinal Chemistry - Anti-Cancer Agents) 8, 92–121.

Collman, J. P., Devaraj, N. K., Eberspacher, T. P. A., Chidsey, C. E. D., 2006. Mixed azide-terminated monolayers: A platform for modifying electrode surfaces. Langmuir 22 (6), 2457–2464.

Cornell, R. M., Schwertmann, U., 2003. The Iron Oxides - Structure, Properties, Reactions, Occurrences and Uses. Wiley-VCH, Weinheim, Germany.

Cote, L. J., Teja, A. S., Wilkinson, A. P., Zhang, Z. J., 2003. Continuous hydrothermal synthesis of $CoFe_2O_4$ nanoparticles. Fluid Phase Equilibria 210 (2), 307–317.

Das, R. S., Agrawal, Y., 2011. Raman spectroscopy: Recent advancements, techniques and applications. Vibrational Spectroscopy 57 (2), 163 – 176.

Das, S., Hendry, M. J., 2011. Application of Raman spectroscopy to identify iron minerals commonly found in mine wastes. Chemical Geology 290 (3-4), 101–108.

de la Fuente, J. M., Barrientos, A. G., Rojas, T. C., Rojo, J., Cañada, J., Fernández, A., Penadés, S., 2001. Gold glyconanoparticles as water-soluble polyvalent models to study carbohydrate interactions. Angewandte Chemie International Edition 40 (12), 2257–2261.

de la Fuente, J. M., Penades, S., 2006. Glyconanoparticles: Types, synthesis and applications in glycoscience, biomedicine and material science. Biochimica et Biophysica Acta 1760 (4), 636–651.

Deul, R., Franck, E. U., 1991. The static dielectric constant of the water-benzene mixture system to 400°C and 2800 bar. Berichte der Bunsen-Gesellschaft für physikalische Chemie 95 (8), 847 – 853.

Eckert, C. A., Knutson, B. L., Debenedetti, P. G., 1996. Supercritical fluids as solvents for chemical and materials processing. Nature 383 (6598), 313–318.

Eisberg, R., Resnick, R., 1985. Quantum Physics of Atoms, Molecules, Solids, Nuclei and Particles, 2nd Edition. Wiley.

El-Boubbou, K., Gruden, C., Huang, X., 2007. Magnetic glyco-nanoparticles: A unique tool for rapid pathogen detection, decontamination, and strain differentiation. Journal of the American Chemical Society 129 (44), 13392–13393.

El Mendili, Y., Bardeau, J.-F., Randrianantoandro, N., Gourbil, A., Greneche, J.-M., Mercier, A.-M., Grasset, F., 2011. New evidences of in situ laser irradiation effects on $\gamma$-$Fe_2O_3$ nanoparticles: A Raman spectroscopic study. Journal of Raman Spectroscopy 42 (2), 239–242.

Finger, L. W., Hazen, R. M., 1980. Crystal structure of isothermal compression of $Fe_2O_3$, $Cr_2O_3$, $V_2O_3$ to 50 kbars. Journal of Applied Physics 51 (10), 5362–5367.

Garcia, I., Marradi, M., Penades, S., 2010. Glyconanoparticles: multifunctional nanomaterials for biomedical applications. Nanomedicine 5 (5), 777–792.

Gatelli, M., Bongiorno, G., Podesta, A., Barbaroni, E., 2004. Characterization procedures of nanomaterials. Tech. rep., INFM UdR Minao Universita, Milano, Italy.

Glasscock, J. A., Barnes, P. R. F., Plumb, I. C., Bendavid, A., Martin, P. J., 2008. Structural, optical and electrical properties of undoped polycrystalline hematite thin films produced using filtered arc deposition. Thin Solid Films 516 (8), 1716–1724.

Goldmann, A. S., Schödel, C., Walther, A., Yuan, J., Loos, K., Müller, A. H. E., 2010. Biomimetic mussel adhesive inspired clickable anchors applied to the functionalization of $Fe_3O_4$ nanoparticles. Macromolecular Rapid Communications 31 (18), 1608–1615.

Gremlich, H.-U., 2000. Infrared and raman spectroscopy. Ullmann's Encyclopedia of Industrial Chemistry.

Gruar, R. I., Tighe, C. J., Darr, J. A., 2013. Scaling-up a confined jet reactor for the continuous hydrothermal manufacture of nanomaterials. Industrial & Engineering Chemistry Research 52 (15), 5270–5281.

Gupta, A. K., Gupta, M., 2005. Synthesis and surface engineering of iron oxide nanoparticles for biomedical applications. Biomaterials 26 (18), 3995–4021.

Haavik, C., Stø len, S., Fjellvåg, H., Hanfland, M., Häusermann, D., 2000. Equation of state of magnetite and its high-pressure modification: Thermodynamics of the Fe-O system at high pressure. American Mineralogist 85, 514–523.

Hakuta, Y., Adschiri, T., Suzuki, T., Chida, T., Seino, K., Arai, K., 1998a. Flow method for rapidly producing barium hexaferrite particles in supercritical water. Journal of the American Ceramic Society 81 (9), 2461–2464.

Hakuta, Y., Haganuma, T., Sue, K., Adschiri, T., Arai, K., 2003. Continuous production of phosphor YAG:Tb nanoparticles by hydrothermal synthesis in supercritical water. Materials Research Bulletin 38 (7), 1257–1265.

Hakuta, Y., Ohashi, T., Hayashi, H., Arai, K., 2004. Hydrothermal synthesis of zirconia nanocrystals in supercritical water. Journal of Materials Research 19 (8), 2230–2234.

Hakuta, Y., Onai, S., Terayama, S., Adschiri, T., Arai, K., 1998b. Production of ultrafine ceria particles by hydrothermal synthesis under supercritical conditions. Journal of Materials Science Letters 17 (14), 1211–1213.

Hakuta, Y., Sue, K.and Adschiri, T., Arai, K., 2001. Understanding of reaction equilibria in supercritical aqueous solutions and development of continous supercritical hydrothermal synthesis processes. In: The 2nd International Meeting on High Pressure Chemical Engineering. Hamburg, Germany, VDI-Gesellschaft Verfahrenstechnik und Chemieingenieurwesen.

Hakuta, Y., Ura, H., Hayashi, H., Arai, K., 2005. Continuous production of $BaTiO_3$ nanoparticles by hydrothermal synthesis. Industrial & Engineering Chemistry Research 44 (4), 840–846.

Hanesch, M., 2009. Raman spectroscopy of iron oxides and (oxy)hydroxides at low laser power and possible applications in environmental magnetic studies. Geophysical Journal International 177 (3), 941–948.

Hao, Y., Teja, A. S., 2003. Continuous hydrothermal crystallization of $\alpha$-$Fe_2o_3$ and $Co_3O_4$ nanoparticles. Journal of Materials Research 18 (2), 415–422.

Hayashi, H., Hakuta, Y., 2010. Hydrothermal synthesis of metal oxide nanoparticles in supercritical water. Materials 3 (7), 3794–3817.

Hayashi, H., Noguchi, T., Islam, N. M., Hakuta, Y., Imai, Y., Ueno, N., 2010a. Hydrothermal synthesis of $BaTiO_3$ nanoparticles using a supercritical continuous flow reaction system. Journal of Crystal Growth 312 (12-13), 1968–1972.

Hayashi, H., Noguchi, T., Islam, N. M., Hakuta, Y., Imai, Y., Ueno, N., 2010b. Hydrothermal synthesis of organic hybrid $BaTiO_3$ nanoparticles using a supercritical continuous flow reaction system. Journal of Crystal Growth 312 (24), 3613–3618.

Hayashi, H., Suino, A., Shimoyama, K., Takesue, M., Tooyama, S., Jr., R. L. S., 2013. Continuous hydrothermal synthesis of $ZnGa_2O_4$:$Mn^{2+}$ nanoparticles at temperatures of 300-500°Cand pressures of 25-35 MPa. The Journal of Supercritical Fluids 77, 1 – 6.

Hayashi, H., Ueda, A., Suino, A., Hiro, K., Hakuta, Y., 2009. Hydrothermal synthesis of yttria stabilized $ZrO_2$ nanoparticles in subcritical and supercritical water using a flow reaction system. Journal of Solid State Chemistry 182 (11), 2985–2990.

Heger, K., Uematsu, M., Franck, E. U., 1980. The static dielectric constant of water at high pressure and temperatures to 500 MPa and 550°C. Berichte der Bunsen-Gesellschaft für physikalische Chemie 84, 758–762.

Hellwege, K.-H., Hellwege, M. (Eds.), 1962. Landolt-Börnstein: Zahlenwerte und Funktionen aus Physik, Astronomie, Geophysik und Technik, 6. Auflage, II. Band Eigenschaften der Materien in ihren Aggregatzuständen, 8.Teil Optische Konstanten. Springer Berlin-Göttingen-Heidelberg.

Hiemenz, P. C., Rajagopalan, R., 1997. Principles of Colloid and Surface Chemistry. Marcel Dekker.

Hirth, T., Fehrenbacher, U., Schweppe, R., Unkelbach, G., 2006. Platform chemicals, monomers and polymers based on renewable materials by hydrothermal processing.

Hong, S.-A., Kim, S. J., Chung, K. Y., Chun, M.-S., Lee, B. G., Kim, J., 2013. Continuous synthesis of lithium iron phosphate ($LiFePO_4$) nanoparticles in supercritical water: Effect of mixing tee. The Journal of Supercritical Fluids 73, 70 – 79.

IAPWS, 1997. Revised release on the IAPWS formulation 1985 for the viscosity of ordinary water substance. Tech. rep., International Association for the Properties of Water and Steam, Erlangen, Germany.

Israelachvili, J., 1992. Intermolecular & Surface Forces, 2nd Edition. Academic Press.

Jacintho, G. V. M., Corio, P., Rubim, J. C., 2007. Surface-enhanced Raman spectra of magnetic nanoparticles adsorbed on a silver electrode. Journal of Electroanalytical Chemistry 603 (1), 27–34.

Jambor, J. L., Dutrizac, J. E., 1998. Occurrence and constitution of natural and synthetic ferrihydrite, a widespread iron oxyhydroxide. Chem. Rev. 98 (7), 2549–2586.

James, M., Ciampi, S., Darwish, T. A., Hanley, T. L., Sylvester, S. O., Gooding, J. J., 2011. Nanoscale water condensation on click-functionalized self-assembled monolayers. Langmuir 27 (17), 10753–10762.

Jähnke, S., 2000. Anwendung der überkritischen Naßoxidation zur Verwertung der Leiterplatten-Restfraktion. Ph.D. thesis, TU Darmstadt.

Johnson, J. W., Norton, D., 1991. Critical phenomena in hydrothermal systems: state, thermodynamic, electrostatic, and transport properties of $H_2O$ in the critical region. American Journal of Science 291 (6), 541–648.

Jubb, A. M., Allen, H. C., 2010. Vibrational spectroscopic characterization of hematite, maghemite, and magnetite thin films produced by vapor deposition. ACS Applied Materials & Interfaces 2 (10), 2804–2812.

Kamiti, M., Popadowski, S., Remsen, E. E., 2007. Advances in the characterization of particle size destributions of abrasive particles used in CMP. Materials Research Society Symposium Proceedings 991 (0991-C04-03), 1–6.

Kawasaki, S.-I., Sue, K., Ookawara, R., Wakashima, Y., Suzuki, A., Hakuta, Y., Arai, K., 2010. Engineering study of continuous supercritical hydrothermal method using a T-shaped mixer: Experimental synthesis of NiO nanoparticles and CFD simulation. The Journal of Supercritical Fluids 54 (1), 96–102.

Kawasaki, S.-i., Xiuyi, Y., Sue, K., Hakuta, Y., Suzuki, A., Arai, K., 2009. Continuous supercritical hydrothermal synthesis of controlled size and highly crystalline anatase $TiO_2$ nanoparticles. The Journal of Supercritical Fluids 50 (3), 276–282.

Kellici, S., Gong, K., Lin, T., Brown, S., Clark, R. J. H., Vickers, M., Cockcroft, J. K., Middelkoop, V., Barnes, P., Perkins, J. M., Tighe, C. J., Darr, J. A., 2010. High-throughput continuous hydrothermal flow synthesis of Zn-Ce oxides: unprecedented solubility of Zn in the nanoparticle fluorite lattice. Philosophical Transactions of the Royal Society A: Mathematical, Physical and Engineering Sciences 368 (1927), 4331–4349.

Kestin, J., Sengers, J., Kamgar-Parsi, B., Levelt Sengers, J., 1984. Thermophysical properties of fluid $H_2O$. Journal of Physical and Chemical Reference Data 13 (1), 175–183.

Kolb, H. C., Finn, M. G., Sharpless, K. B., 2001. Click chemistry: Diverse chemical function from a few good reactions. Angewandte Chemie International Edition 40 (11), 2004–2021.

Kriete, A., Gundlach, H., Amelinckx, S., Reimer, L., 2005. Microscopy. Ullmann's Encyclopedia of Industrial Chemistry.

Kruse, A., Dinjus, E., 2007. Hot compressed water as reaction medium and reactant: Properties and synthesis reactions. The Journal of Supercritical Fluids 39 (3), 362–380.

Lagaly, G., 2007. Colloids. In: Ullmann's Encyclopedia of Industrial Chemistry. Wiley.

Laidlaw, I., Steinmetz, M., 2005. Analytical Ultracentrifugation, Techniques and Methods. The Royal Society of Chemistry, Ch. 14: Introduction to Differential Sedimentation, pp. 270–290.

Laidler, K. J., Meiser, J. H., 1995. Physical Chemistry, 2nd Edition. Houghton Mifflin, Boston.

Lallana, E., Sousa-Herves, A., Fernandez-Trillo, F., Riguera, R., Fernandez-Megia, E., 2012. Click chemistry for drug delivery nanosystems. Pharmaceutical Research 29, 1–34.

Lam, U. T., Mammucari, R., Suzuki, K., Foster, N. R., 2008. Processing of iron oxide nanoparticles by supercritical fluids. Industrial & Engineering Chemistry Research 47 (3), 599–614.

Lepenies, B., Yin, J., Seeberger, P. H., 2010. Applications of synthetic carbohydrates to chemical biology. Current Opinion in Chemical Biology 14 (3), 404–411.

Lester, E., Aksomaityte, G., Li, J., Gomez, S., Gonzalez-Gonzalez, J., Poliakoff, M., 2012. Controlled continuous hydrothermal synthesis of cobalt oxide ($Co_3O_4$) nanoparticles. Progress in Crystal Growth and Characterization of Materials 58 (1), 3 – 13.

Lester, E., Blood, P., Denyer, J., Giddings, D., Azzopardi, B., Poliakoff, M., 2006. Reaction engineering: The supercritical water hydrothermal synthesis of nano-particles. The Journal of Supercritical Fluids 37 (2), 209–214.

Lester, E., Blood, P. J., Denyer, J. P., Azzopardi, B. J., Li, J., Poliakoff, M., 2010. Impact of reactor geometry on continuous hydrothermal synthesis mixing. Materials Research Innovations 14, 19–26.

Li, N., Binder, W. H., 2011. Click-chemistry for nanoparticle-modification. Journal of Materials Chemistry 21 (42), 16717–16734.

Lin, T., Kellici, S., Gong, K., Thompson, K., Darr, J. A., 2010. The rapid automated materials synthesis instrument (RAMSI): a high throughput combinatorial robot for nanoceramics discovery. Advances in Science and Technology 62, 215–20.

Longtin, D. R., Shettle, E. P., Hummel, J. R., Pryce, J. F., 1988. A wind desert aerosol model: Radiative properties. Scientific Report No. 6, Air Force Geophysics Laboratory, Air Force Systems Command, United States Air Force.

Lu, J., Hakuta, Y., Hayashi, H., Ohashi, T., Nagase, T., Hoshi, Y., Sato, K., Nishioka, M., Inoue, T., Hamakawa, S., 2008. Preparation of $Ca_{0.8}Sr_{0.2}Ti_{1-x}Fe_xO_{3-d}$ (x=0.1-0.3) nanoparticles using a flow supercritical reaction system. The Journal of Supercritical Fluids 46 (1), 77–82.

Lu, J., Minami, K., Takami, S., Adschiri, T., 2013. Rapid and continuous synthesis of cobalt aluminate nanoparticles under subcritical hydrothermal conditions with in-situ surface modification. Chemical Engineering Science 85, 50 – 54.

Lu, J., Minami, K., Takami, S., Shibata, M., Kaneko, Y., Adschiri, T., 2012. Supercritical hydrothermal synthesis and in situ organic modification of indium tin oxide nanoparticles using continuous-flow reaction system. ACS Applied Materials & Interfaces 4 (1), 351–354.

Lu, X., Sun, F., Wang, J., Zhong, J., Dong, Q., 2009. A facile route to prepare organic/inorganic hybrid nanomaterials by "click chemistry". Macromolecular Rapid Communications 30 (24), 2116–2120, 1521-3927.

Lübbe, M., Gigler, A. M., Stark, R. W., Moritz, W., 2010. Identification of iron oxide phases in thin films grown on $Al_2O_3$(0 0 0 1) by Raman spectroscopy and X-ray diffraction. Surface Science. 604 (7-8), 679–685.

Luemmen, N., Kvamme, B., 2007. Kinetics of nacl nucleation in supercritical water investigated by molecular dynamics simulations. Physical Chemistry Chemical Physics 9 (25), 3251–3260.

Luth, H., 2001. Solid Surfaces, Interfaces, and Thin FIlms, 4th Edition. Springer.

Lutz, J.-F. c., Zarafshani, Z., 2008. Efficient construction of therapeutics, bioconjugates, biomaterials and bioactive surfaces using azide-alkyne "click" chemistry. Advanced Drug Delivery Reviews 60 (9), 958–970.

Mae, K., Suzuki, A., Maki, T., Hakuta, Y., Sato, H., Arai, K., 2007. A new micromixer with needle adjustment for instant mixing and heating under high pressure and high temperature. Journal of Chemical Engineering of Japan 40 (12), 1101–1107.

Malvern Instruments, 2012a. Electrophoretic light scattering. Accessed 27 December 2012.
URL `http://www.malvern.com/labeng/technology/electrophoretic-light-scattering.htm`

Malvern Instruments, 2012b. Zeta potential measurement using laser Doppler electrophoresis (LDE). Accessed 27 December 2012.
URL `http://www.malvern.com/labeng/technology/zeta_potential/zeta_potential_LDE.htm`

Marradi, M., Martín-Lomas, M., Penadés, S., 2010. Glyconanoparticles: Polyvalent tools to study carbohydrate-based interactions. In: Horton, D. (Ed.), Advances in Carbohydrate Chemistry and Biochemistry. Vol. 64 of Advances in Carbohydrate Chemistry and Biochemistry. Academic Press, pp. 211 – 290.

Marrone, P. A., Hodes, M., Smith, K. A., Tester, J. W., 2004. Salt precipitation and scale control in supercritical water oxidation–part B: commercial/full-scale applications. The Journal of Supercritical Fluids 29 (3), 289–312.

Marshall, W. L., Franck, E. U., 1981. Ion product of water substance, 0-1000°C, 1-10,000 bars - new international formulation and its background. Journal of Physical and Chemical Reference Data 10 (2), 295–304.

Martínez-Ávila, O., Hijazi, K., Marradi, M., Clavel, C., Campion, C., Kelly, C., Penadés, S., 2009. Gold manno-glyconanoparticles: Multivalent systems to block HIV-1 gp120 binding to the lectin DC-SIGN. Chemistry - A European Journal 15 (38), 9874–9888.

Matson, D. W., Linehan, J. C., Bean, R. M., 1992. Ultrafine iron oxide powders generated using a flow-through hydrothermal process. Materials Letters 14 (4), 222–226.

Mazzetti, L., Thistlethwaite, P. J., 2002. Raman spectra and thermal transformations of ferrihydrite and schwertmannite. Journal of Raman Spectroscopy 33 (2), 104–111.

Michel, F. M., Ehm, L., Antao, S. M., Lee, P. L., Chupas, P. J., Liu, G., Strongin, D. R., Schoonen, M. A. A., Phillips, B. L., Parise, J. B., 2007. The structure of ferrihydrite, a nanocrystalline material. Science 316 (5832), 1726–1729.

Middelkoop, V., Boldrin, P., Peel, M., Buslaps, T., Barnes, P., Darr, J. A., Jacques, S. D. M., 2009. Imaging the inside of a continuous nanoceramic synthesizer under supercritical water conditions using high-energy synchrotron X-radiation. Chemistry of Materials 21 (12), 2430–2435.

Mousavand, T., Ohara, S., Umetsu, M., Zhang, J., Takami, S., Naka, T., Adschiri, T., 2007. Hydrothermal synthesis and in situ surface modification of boehmite nanoparticles in supercritical water. The Journal of Supercritical Fluids 40 (3), 397–401.

Mousavand, T., Takami, S., Umetsu, M., Ohara, S., Adschiri, T., 2006. Supercritical hydrothermal synthesis of organic-inorganic hybrid nanoparticles. Journal of Materials Science 41 (5), 1445–1448.

Müller, R., Schuhmann, R., 1997. Teilchengrössenmessungen in der Laborpraxis. Wissenschaftliche Verlagsgesellschaft mbH Stuttgart.

Nadler, M., Mahrholz, T., Riedel, U., Schilde, C., Kwade, A., 2008. Preparation of colloidal carbon nanotube dispersions and their characterisation using a disc centrifuge. Carbon 46 (11), 1384–1392.

Nahtigal, I. G., Zasetsky, A. Y., Svishchev, I. M., 2008. Nucleation of nacl nanoparticles in supercritical water: Molecular dynamics simulations. Journal of Physical Chemistry B 112 (25), 7537–7543.

Nebhani, L., Barner-Kowollik, C., 2009. Orthogonal transformations on solid substrates: Efficient avenues to surface modification. Advanced Materials 21 (34), 3442–3468.

Noguchi, T., Matsui, K., Islam, N. M., Hakuta, Y., Hayashi, H., 2008. Rapid synthesis of $\gamma$-$Al_2O_3$ nanoparticles in supercritical water by continuous hydrothermal flow reaction system. The Journal of Supercritical Fluids 46 (2), 129–136.

Nunney, T., White, R., 2011. Characterizing materials for energy generation using X-ray photoelectron spectroscopy (XPS). Microscopy Today 19 (2), 22–28.

Ojeda, R., de Paz, J. L., Barrientos, A. G., Martín-Lomas, M., Penadés, S., 2007. Preparation of multifunctional glyconanoparticles as a platform for potential carbohydrate-based anticancer vaccines. Carbohydrate Research 342 (3-4), 448–459.

Parry, K. L., Shard, A. G., Short, R. D., White, R. G., Whittle, J. D., Wright, A., 2006. ARXPS characterisation of plasma polymerised surface chemical gradients. Surface and Interface Analysis 38 (11), 1497–1504.

Paulus, E. F., Gieren, A., 2001. Structure analysis by diffraction. Ullmann's Encyclopedia of Industrial Chemistry.

Peak, D., Regier, T., 2012. Direct observation of tetrahedrally coordinated Fe(III) in ferrihydrite. Environmental Science & Technology 46 (6), 3163–3168.

Pecharromán, C., González-Carreño, T., Iglesias, J. E., 1995. The infrared dielectric properties of maghemite, $\gamma$-$Fe_2O_3$, from reflectance measurement on pressed powders. Physics and Chemistry of Minerals 22 (1), 21–29.

Peterson, A. A., Vontobel, P., Vogel, F., Tester, J. W., 2008. In situ visualization of the performance of a supercritical-water salt separator using neutron radiography. The Journal of Supercritical Fluids 43 (3), 490–499.

Popović, Z., Dohčević-Mitrović, Z., Šćepanović, M., Grujić-Brojčin, M., Aškrabić, S., 2011. Raman scattering on nanomaterials and nanostructures. Annalen der Physik 523 (1-2), 62–74.

Portella, J., Mateos, D., Mancini, F., Marraud, C., Cansell, F., 2007. Hydrothermal oxidation with multi-injection of oxygen: Simulation and experimental data. The Journal of Supercritical Fluids 40 (2), 258–262.

Prikopsky, K., Wellig, B., von Rohr, P. R., 2007. SCWO of salt containing artificial wastewater using a transpiring-wall reactor: Experimental results. The Journal of Supercritical Fluids 40 (2), 246–257.

R Development Core Team, 2011. R: A Language and Environment for Statistical Computing. R Foundation for Statistical Computing, Vienna, Austria, ISBN 3-900051-07-0.
URL `http://www.R-project.org/`

Rangappa, D., Ohara, S., Umetsu, M., Naka, T., Adschiri, T., 2008. Synthesis, characterization and organic modification of copper manganese oxide nanocrystals under supercritical water. The Journal of Supercritical Fluids 44 (3), 441–445.

Reid, E., Cooney, R., Hendra, P., Fleischmann, M., 1977. A Raman spectroscopic study of corrosion of lead electrodes in aqueous chloride media. Journal of Electroanalytical Chemistry 80, 405–408.

Riedel, E., 1990. Anorganische Chemie. deGruyter.

Rivière, J. C., Arlinghaus, H. F., Hutter, H., Jenett, H., Bauer, P., Palmetshofer, L., 2011. Surface and thin-film analysis: 2. Electron detection. In: Ullmann's Encyclopedia of Industrial Chemistry. Wiley.

Rogalinski, T., Liu, K., Albrecht, T., Brunner, G., 2008. Hydrolysis kinetics of biopolymers in subcritical water. The Journal of Supercritical Fluids 46 (3), 335–341.

Rutledge, R. D., Warner, C. L., Pittman, J. W., Addleman, R. S., Engelhard, M., Chouyyok, W., Warner, M. G., 2010. Thiol-ene induced diphosphonic acid functionalization of superparamagnetic iron oxide nanoparticles. Langmuir 26 (14), 12285–12292.

Sahraneshin, A., Takami, S., Hojo, D., Minami, K., Arita, T., Adschiri, T., 2012a. Synthesis of shape-controlled and organic-hybridized hafnium oxide nanoparticles under sub- and supercritical hydrothermal conditions. The Journal of Supercritical Fluids 62, 190 – 196.

Sahraneshin, A., Takami, S., Minami, K., Hojo, D., Arita, T., Adschiri, T., 2012b. Synthesis and morphology control of surface functionalized nanoscale yttrium aluminum garnet particles via supercritical hydrothermal method. Progress in Crystal Growth and Characterization of Materials 58 (1), 43 – 50.

Sanchez, C., Belleville, P., Popall, M., Nicole, L., 2011. Applications of advanced hybrid organic-inorganic nanomaterials: from laboratory to market. Chemical Society Reviews 40 (2), 696–753.

Schmitt, M., Popp, J., 2006. Raman spectroscopy at the beginning of the twenty-first century. Journal of Raman Spectroscopy 37 (1-3), 20–28.

Schrader, B., 1983. Ramanspektrometrie. Fresenius Zeitschrift für Analytische Chemie 314, 363–382.

Schultz, N., 2007. Application of magnetic separation technology for the recovery and re-use of immobilised lipase of candida antarctica A-type (CALA). Ph.D. thesis, Universität Fredericiana Karlsruhe (TH).

Schwertmann, U., Cornell, R. M., 2000. Iron Oxides in the Laboratory - Preparation and Characterization, 2nd Edition. Wiley-VCH, Weinheim, Germany.

Scofield, J. H., 1976. Hartree-Slater subshell photoionization cross-sections at 1254 and 1487 eV. Journal of Electron Spectroscopy and Related Phenomena 8 (2), 129–137.

Serres, S., Anthony, D. C., Jiang, Y., Broom, K. A., Campbell, S. J., Tyler, D. J., van Kasteren, S. I., Davis, B. G., Sibson, N. R., 2009. Systemic inflammatory response reactivates immune-mediated lesions in rat brain. Journal of Neuroscience 29 (15), 4820–4828.

Serway, R. A., 1996. Physics for Scientists and Engineers, 4th Edition. Saunders College Publishing, Philadelphia.

Shackelford, J. F., 1985. Introduction to Materials Science for Engineers. MacMillan.

Shock, E. L., Oelkers, E. H., Johnson, J. W., Sverjensky, D. A., Helgeson, H. C., 1992. Calculation of the thermodynamic properties of aqueous species at high pressures and temperatures: effective electrostatic radii, dissociation constants and standard partial molal properties to 1000°C and 5 kbar. Journal of the Chemical Society, Faraday Transactions 88 (6), 803–26.

Sierra-Pallares, J., Marchisio, D. L., Alonso, E., Parra-Santos, M. T., Castro, F., José Cocero, M., 2011. Quantification of mixing efficiency in turbulent supercritical water hydrothermal reactors. Chemical Engineering Science 66 (8), 1576–1589.

Sihelníková, L., Tvaroška, I., 2007. Step by step towards understanding gold glyconanoparticles as elements of the nanoworld. Chemical Papers 61 (4), 237–255.

Smith, S. J., Page, K., Kim, H., Campbell, B. J., Boerio-Goates, J., Woodfield, B. F., 2012. Novel synthesis and structural analysis of ferrihydrite. Inorganic Chemistry 51, 6421–6424.

Sousa, M. H., Tourinho, F. A., Rubim, J. C., 2000. Use of Raman micro-spectroscopy in the characterization of $M^{II}Fe_2O_4$ (M = Fe, Zn) electric double layer ferrofluids. Journal of Raman Spectroscopy 31 (3), 185–191.

Such, G. K., Johnston, A. P., Liang, K., Caruso, F., 2012. Synthesis and functionalization of nanoengineered materials using click chemistry. Progress in Polymer Science 37 (7), 985 – 1003.

Sue, K., Kawasaki, S.-i., Suzuki, M., Hakuta, Y., Hayashi, H., Arai, K., Takebayashi, Y., Yoda, S., Furuya, T., 2011. Continuous hydrothermal synthesis of $Fe_2O_3$, NiO, and CuO nanoparticles by superrapid heating using a T-type micro mixer at 673 K and 30 MPa. Chemical Engineering Journal 166 (3), 947–953.

Sue, K., Murata, K., Kimura, K., Arai, K., 2003. Continuous synthesis of zinc oxide nanoparticles in supercritical water. Green Chemistry 5 (5), 659–662.

Sue, K., Sato, T., Kawasaki, S.-i., Takebayashi, Y., Yoda, S., Furuya, T., Hiaki, T., 2010. Continuous hydrothermal synthesis of $Fe_2O_3$ nanoparticles using a central collision-type micromixer for rapid and homogeneous nucleation at 673 K and 30 MPa. Industrial & Engineering Chemistry Research 49 (18), 8841–8846.

Sue, K., Suzuki, M., Arai, K., Ohashi, T., Ura, H., Matsui, K., Hakuta, Y., Hayashi, H., Watanabe, M., Hiaki, T., 2006. Size-controlled synthesis of metal oxide nanoparticles with a flow-through supercritical water method. Green Chemistry 8 (7), 634–638.

Taguchi, M., Takami, S., Adschiri, T., Nakane, T., Sato, K., Naka, T., 2011. Supercritical hydrothermal synthesis of hydrophilic polymer-modified water-dispersible $CeO_2$ nanoparticles. CrystEngComm 13 (8), 2841–2848.

Taguchi, M., Takami, S., Adschiri, T., Nakane, T., Sato, K., Naka, T., 2012. Synthesis of surface-modified monoclinic $ZrO_2$ nanoparticles using supercritical water. CrystEngComm 14, 2132–2138.

Takami, S., Ohara, S., Adschiri, T., Wakayama, Y., Chikyow, T., 2008. Continuous synthesis of organic-inorganic hybridized cubic nanoassemblies of octahedral cerium oxide nanocrystals and hexanedioic acid. Dalton Transactions (40), 5442–5446.

Tanuma, S., Powell, C. J., Penn, D. R., 1994. Calculations of electron inelastic mean free paths. V. data for 14 organic compounds over the 50–2000 eV range. Surface and Interface Analysis 21 (3), 165–176.

Taylor, M. E., Drickamer, K., 2003. Introduction to Glycobiology. University Press, Oxford.

Teja, A. S., Koh, P.-Y., 2009. Synthesis, properties, and applications of magnetic iron oxide nanoparticles. Progress in Crystal Growth and Characterization of Materials 55 (1-2), 22–45.

Tepper, T., Ross, C. A., Dionne, G. F., 2004. Microstructure and optical properties of pulsed-laser-deposited iron oxide films. IEEE Transactions on Magnetics 40 (3), 1685–1690.

Thibeau, R. J., Brown, C. W., Heidersbach, R. H., 1978. Raman spectra of possible corrosion products of iron. Applied Spectroscopy 32 (6), 532–535.

Thomason, T. B., Modell, M., 1984. Supercritcal water destruction of aqueous wastes. Hazardous Waste 1 (4), 453–467.

Togashi, T., Naka, T., Asahina, S., Sato, K., Takami, S., Adschiri, T., 2011. Surfactant-assisted one-pot synthesis of superparamagnetic magnetite nanoparticle clusters with tunable cluster size and magnetic field sensitivity. Dalton Transactions 40 (5), 1073–1078.

Tyrsted, C., Becker, J., Hald, P., Bremholm, M., Pedersen, J. S., Chevallier, J., Cerenius, Y., Iversen, S. B., Iversen, B. B., 2010. In-situ synchrotron radiation study of formation and growth of crystalline $Ce_xZr_{1-x}O_2$ nanoparticles synthesized in supercritical water. Chemistry of Materials 22 (5), 1814–1820.

Tüysüz, H., Salabaş, E. L., Weidenthaler, C., Schüth, F., 2008. Synthesis and magnetic investigation of ordered mesoporous two-line ferrihydrite. Journal of the American Chemical Society 130 (1), 280–287.

van Kasteren, S. I., Campbell, S. J., Serres, S., Anthony, D. C., Sibson, N. R., Davis, B. G., 2009. Glyconanoparticles allow pre-symptomatic in vivo imaging of brain disease. Proceedings of the National Academy of Sciences 106 (1), 18–23.

Van Wylen, G., Sonntag, R., Borgnakke, C., 1994. Fundamentals of Classical Thermodynamics. Wiley, New York.

Vangelista, S., Mantovan, R., Cocco, S., Lamperti, A., Salicio, O., Fanciulli, M., 2012. Chemical vapor deposition growth of $Fe_3O_4$ thin films and $Fe/Fe_3O_4$ bi-layers for their integration in magnetic tunnel junctions. Thin Solid Films 520 (14), 4617 – 4621.

Wagner, W., Pruss, A., 2002. The IAPWS formulation 1995 for the thermodynamic properties of ordinary water substance for general and scientific use. Journal of Physical and Chemical Reference Data 31 (2), 387–535.

Wakashima, Y., Suzuki, A., Kawasaki, S.-i., Matsui, K., Yukiya, H., 2007. Development of a new swirling micro mixer for continuous hydrothermal synthesis of nano-size particles. Journal of Chemical Engineering of Japan 40 (8), 622–629.

Wang, J., White, W. B., Adair, J. H., 2005. Optical properties of hydrothermally synthesized hematite particulate pigments. Journal of the American Ceramic Society 88 (12), 3449–3454.

Wang, X., Ramström, O., Yan, M., 2010. Glyconanomaterials: Synthesis, characterization, and ligand presentation. Advanced Materials 22 (17), 1946–1953.

Weingärtner, H., Franck, E. U., 2005. Überkritisches Wasser als Lösungsmittel. Angewandte Chemie 117 (18), 2730–2752.

Weng, X., Cockcroft, J. K., Hyett, G., Vickers, M., Boldrin, P., Tang, C. C., Thompson, S. P., Parker, J. E., Knowles, J. C., Rehman, I., Parkin, I., Evans, J. R. G., Darr, J. A.,

2009. High-throughput continuous hydrothermal synthesis of an entire nanoceramic phase diagram. Journal of Combinatorial Chemistry 11 (5), 829–834.

White, M. A., Johnson, J. A., Koberstein, J. T., Turro, N. J., 2007. Toward the syntheses of universal ligands for metal oxide surfaces: Controlling surface functionality through click chemistry. Journal of the American Chemical Society 129 (14), 4504–4504.

Wollman, E. W., Kang, D., Frisbie, C. D., Lorkovic, I. M., Wrighton, M. S., 1994. Photosensitive self-assembled monolayers on gold: Photochemistry of surface-confined aryl azide and cyclopentadienylmanganese tricarbonyl. Journal of the American Chemical Society 116 (10), 4395–4404.

Xu, C., Teja, A. S., 2008. Continuous hydrothermal synthesis of iron oxide and pva-protected iron oxide nanoparticles. The Journal of Supercritical Fluids 44 (1), 85–91.

Yin, Y., Alivisatos, A. P., 2005. Colloidal nanocrystal synthesis and the organic-inorganic interface. Nature 437 (7059), 664–670.

Yu, M. K., Jeong, Y. Y., Park, J., Park, S., Woong, J., Jung, K., Min, J., Kim, K., Jon, S., 2008. Drug-loaded superparamagnetic iron oxide nanoparticles for combined cancer imaging and therapy in vivo. Angewandte Chemie, International Edition in English 47 (29), 5362–5365.

Zhang, J., Ohara, S., Umetsu, M., Naka, T., Hatakeyama, Y., Adschiri, T., 2007. Colloidal ceria nanocrystals: A tailor-made crystal morphology in supercritical water. Advanced Materials 19 (2), 203–206.

Zhou, L., Yuan, J., Wei, Y., 2011. Core-shell structural iron oxide hybrid nanoparticles: from controlled synthesis to biomedical applications. Journal of Materials Chemistry 21 (9), 2823–2840.

Ziegler, K. J., Doty, R. C., Johnston, K. P., Korgel, B. A., 2001. Synthesis of organic monolayer-stabilized copper nanocrystals in supercritical water. Journal of the American Chemical Society 123 (32), 7797–7803.

# Appendix A

# Acronyms

| | |
|---|---|
| AGM | alternating gradient magnetometer |
| ASS | atomic spectroscopy standard |
| CFD | computational fluid dynamics |
| CuAAC | copper(I)-catalyzed azide-alkyne cycloaddition |
| DCS | differential centrifugal sedimenation |
| DMSO | dimethyl sulfoxide |
| EAL | effective attenuation lengths |
| EDX | energy-dispersive X-ray spectroscopy |
| FTIR | fourier transform infrared spectroscopy |
| GNP | glyconanoparticle |
| HAADF | high angle annular dark field |
| HKF | Helgeson-Kirkham-Flowers (model) |
| HRTEM | high-resolution transmission electron microscopy |
| IAPWS | The International Association for the Properties of Water and Steam |
| ICP-OES | ion coupled plasma optical emission spectrometry |
| IC | ion chromatography |
| I.D. | inner diameter |
| ITO | tin-doped indium oxide |
| LSCM | laser scanning confocal microscopy |

| | |
|---|---|
| MF | membrane filtration |
| MWCO | molecular weight cut-off |
| MRI | magnetic resonance imaging |
| nc-$H_2O$ | near-critical water |
| NIST | National Institute of Standards and Technology |
| NMR | nuclear magnetic resonance |
| O.D. | outer diameter |
| PID | proportional-integral-derivative (controller) |
| PESH | permanent hydrophylized polyethersulfon |
| PAA | polyacrylic acid |
| PDF | powder diffraction file |
| PSD | particle size distribution |
| PVA | polyvinyl alcohol |
| PXRD | powder X-ray diffraction |
| RTDS | rapid thermal decomposition of precursors in solution |
| XPS | X-ray photoelectron spectroscopy |
| sc-$H_2O$ | supercritical water |
| SAXS | small-angle X-ray scattering |
| SCWG | supercritical water gasification |
| SCWO | supercritical water oxidation |
| STEM | scanning transmission electron microscopy |
| TADDI | tomography angle diffraction imaging |
| TEDDI | tomography energy-dispersive diffraction imaging |
| TEM | transmission electron microscopy |
| TGA | thermogravimetric analysis |
| UV | ultraviolet |

| | |
|---|---|
| WAXS | wide-angle X-ray scattering |
| XRD | X-ray diffraction |

# Appendix B

# Nomenclature

## B.1 Latin characters

| Symbol | Meaning | Value/Units |
|---|---|---|
| $a_{\mathrm{r}}$ | frequency factor at reference dielectric constant ($\epsilon_r$) | |
| $c_{\mathrm{salt}}$ | concentration of the metal salt in $\dot{m}_{salt}$ | mol $L^{-1}$ |
| $c_{\mathrm{product}}$ | concentration of the metal oxide in $\dot{m}_{total}$ | g $L^{-1}$ |
| $c_{\mathrm{Fe,\ after\ mix}}$ | concentration of iron after the mxing with the hot compressed water stream | mol $kg^{-1}$ |
| $d$ | interplanar distance (in XRD) | nm |
| $d_{\mathrm{crystal}}$ | mean crystallite size | nm |
| $D$ | Stokes equivalent diameter (in DCS) | $\mu$m |
| $E_{\mathrm{bin}}$ | binding energy of electrons | J |
| $E_{\mathrm{kin}}$ | kinetic energy of ejected photoelectrons | J |
| $E_{\mathrm{r}}$ | activation energy at reference dielectric constant ($\epsilon_r$) | J $mol^{-1}$ |
| $F_{\mathrm{water}}$ | measurement point for mass flow rate of the main water stream | |
| $F_{\mathrm{quench}}$ | measurement point for mass flow rate of the quenching stream | |
| $g$ | gravitational acceleration | m $s^{-2}$ |
| $h$ | sedimentation distance | m |

| Symbol | Meaning | Value/Units |
|---|---|---|
| $h$ | Planck constant | $6.626 \times 10^{-34}$ m$^2$ kg s$^{-1}$ |
| $k$ | reaction rate coefficient | |
| $k$ | extinction coefficient (imaginary part of complex index of refraction) in context of DCS | |
| $K$ | equilibrium constant | |
| $\dot{m}_{salt}$ | mass flow rate of metal salt stream | kg h$^{-1}$ |
| $\dot{m}_{water}$ | mass flow rate of the water stream | kg h$^{-1}$ |
| $\dot{m}_{total}$ | mass flow rate of the combined streams | kg h$^{-1}$ |
| $\dot{m}_{max}$ | maximum mass flow rate of operation of the pilot plant | kg h$^{-1}$ |
| $M_s$ | saturation magnetization | A m$^2$ kg$^{-1}$ |
| $M_r$ | remanence/remanent magnetization | A m$^2$ kg$^{-1}$ |
| $n$ | real part of complex index of refraction | |
| $Q_3$ | cumulative particle size distribution based on mass | |
| $q_3$ | differential particle size distribution based on mass | cm$^{-1}$ |
| $R$ | gas constant | 8.314 J K$^{-1}$ mol$^{-1}$ |
| $S$ | estimated metal oxide solubility under nc-$H_2O$ or sc-$H_2O$ conditions | mol kg$^{-1}$ |
| $S_0$ | metal concentration of the starting solution | mol kg$^{-1}$ |
| $t$ | time | s |
| $T$ | absolute temperature | K |
| $T_0$ | room temperature | K |
| $T_c$ | temperature at the critical point | K |
| $T_{after\,mix}$ | temperature measured directly after the mixing unit | K |
| $T_{reactor}$ | temperature measured in the reaction zone | K |

| Symbol | Meaning | Value/Units |
|---|---|---|
| $T_{cool}$ | temperature measured after cooling | K |
| $T_{hot\,water}$ | temperature of the hot compressed water stream measured directly before the first mixing unit | K |
| $T_{salt}$ | temperature of the aqueous metal salt stream measured directly before the first mixing unit | K |
| $T_{max}$ | maximum temperature of operation of the pilot plant | K |
| $p_c$ | pressure at the critical point | MPa |
| $p_{before\,mix}$ | absolute pressure measured directly before the mixing unit | MPa |
| $p_{after\,mix}$ | absolute pressure measured after the mixing units and directly before the entrance into the cooler | MPa |
| $p_{max}$ | maximum pressure of operation of the pilot plant | K |
| $pH_{salt}$ | pH-value of the starting salt solution | |
| $pH_{product}$ | pH-value of the product dispersion | |
| $R$ | position (radius) in DCS | m |
| $R_0$ | initial sample position (radius) in DCS | m |
| $R_f$ | detector position (radius) in DCS | m |
| $s$ | non-sphericity factor; average aspect ratio (longest to shortest dimension) for all possible orientations of the particles | |
| $x_{k,3}$ | mode of a particle size distribution based on mass | nm |
| $x_{50,3}$ | median of a particle size distribution based on mass | nm |
| $x_{y,3}$ | y% of the particles have a diameter less than or equal to x (based on mass) | nm |
| $x_{g,3}$ | geometric mean | nm |
| $X$ | conversion | |

## B.2 Greek symbols

| Symbol | Meaning | Value/Units |
| --- | --- | --- |
| $\delta_{\mathrm{ppm}}$ | chemical shift (NMR) | ppm |
| $\delta_{\mathrm{w}}$ | wall thickness | mm |
| $\Delta p$ | pressure difference between $p_{\mathrm{after\ cool}}$ and $p_{\mathrm{before\ mix}}$ | |
| $\epsilon$ | dielectric constant | |
| $\epsilon_{\mathrm{r}}$ | reference dielectric constant | |
| $\eta$ | viscosity | $\mathrm{kg\ s^{-1}\ m^{-1}}$ |
| $\kappa$ | dispersity according to VDI guidline 3491 | |
| $\lambda$ | wavelength | nm |
| $\rho$ | density | $\mathrm{kg\ m^{-3}}$ |
| $\rho_{\mathrm{c}}$ | density at the critical point | $\mathrm{kg\ m^{-3}}$ |
| $\rho_0$ | density at ambient conditions | $\mathrm{kg\ m^{-3}}$ |
| $\rho_{\mathrm{p}}$ | particle density | $\mathrm{kg\ m^{-3}}$ |
| $\rho_{\mathrm{f}}$ | fluid density | $\mathrm{kg\ m^{-3}}$ |
| $\rho_{\mathrm{net}}$ | density | $\mathrm{kg\ m^{-3}}$ |
| $\phi_{\mathrm{m}}$ | mixing value defined as the ratio between $\dot{m}_{\mathrm{salt}}$ and $\dot{m}_{\mathrm{total}}$ | |
| $\Phi_{\mathrm{S}}$ | spectrometer work function (in XPS) | J |
| $\Psi$ | constant determined for the specific reaction system (for $k$) | $\mathrm{J\ mol^{-1}}$ |
| $\sigma_{\mathrm{g,3}}$ | geometric standard deviation | |
| $\sigma$ | supersaturation | |
| $\tau$ | residence time | s |
| $\theta$ | angle between the incident waves and atomic planes (in XRD) | |

| Symbol | Meaning | Value/Units |
|---|---|---|
| $\omega$ | angular velocity | rad $s^{-1}$ |
| $\omega_j$ | constant determined for the specific reaction system (for $K$) | J $mol^{-1}$ |

# Appendix C

# Additional data for Chapter 4

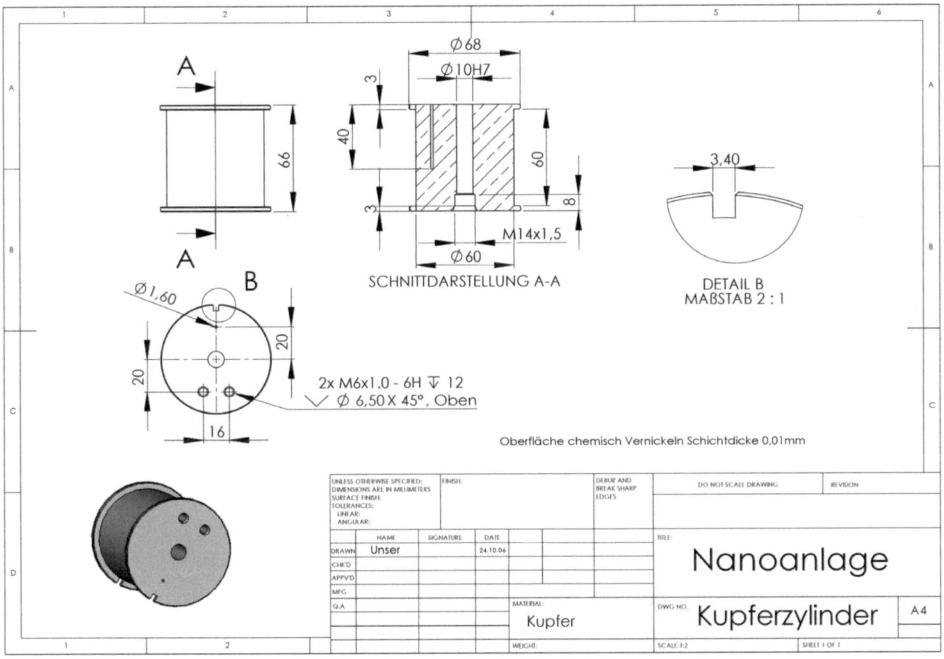


**Figure C.1:** Engineering drawing of heating blocks for preheating the main water stream and maintaining the reactor temperature.

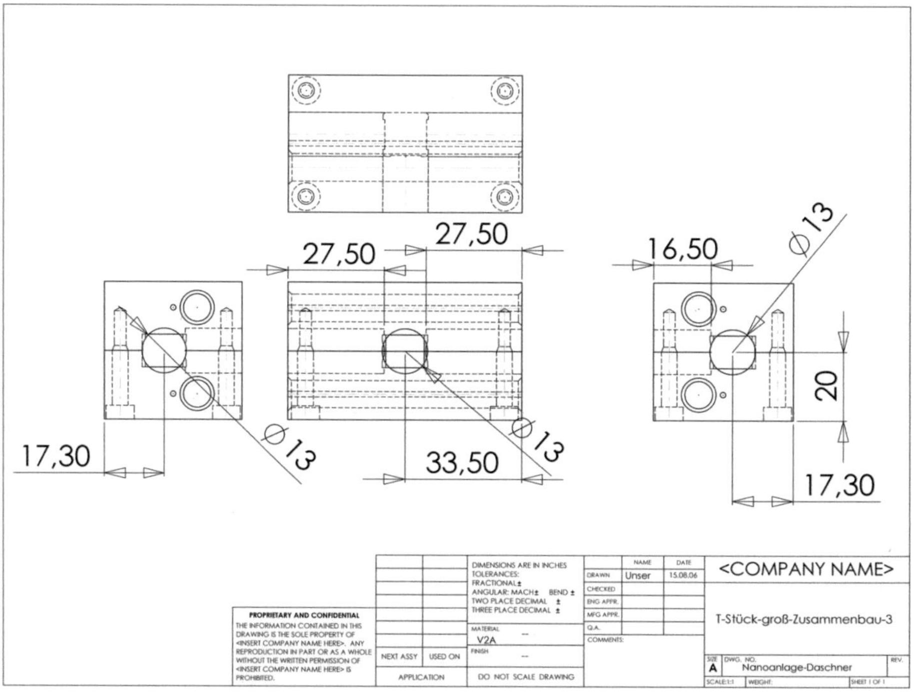


**Figure C.2:** Engineering drawing of heating element for a 1/8 in. T-union allowing the introduction of a thermocouple into the stream.

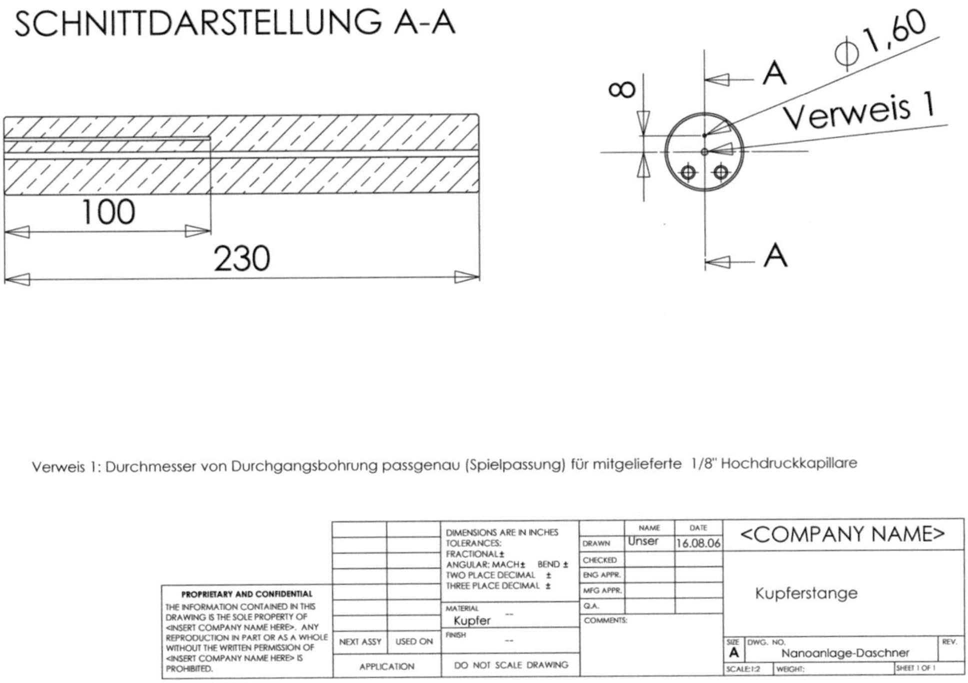


**Figure C.3:** Engineering drawing of heating element for a 1/8 in. reaction capillary.

**Figure C.4 *(facing page)*:** Flow diagram of the high pressure pilot plant in one variant. Refer to Chapter 4 for details on the specific unit operations and the different setups.

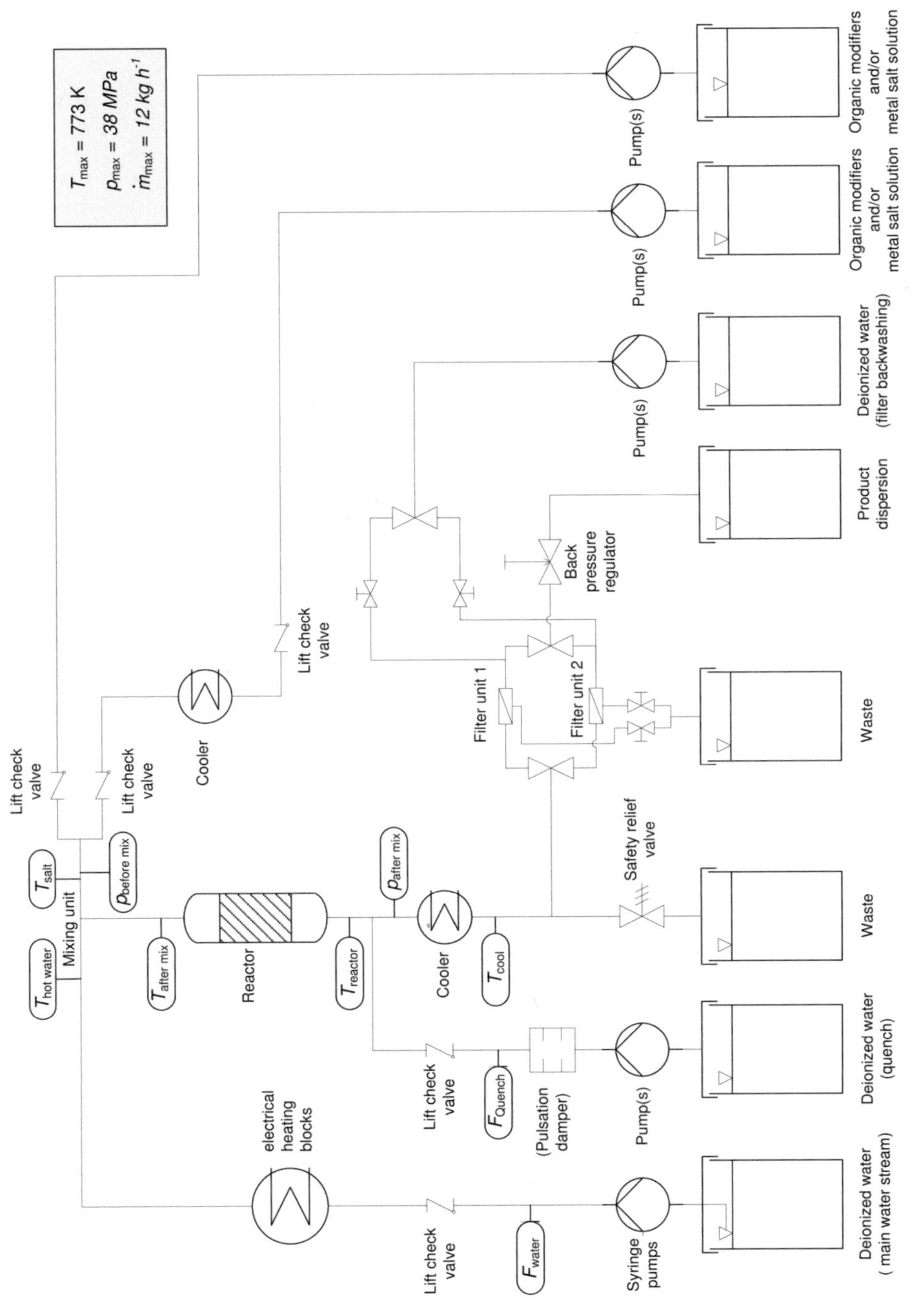

$T_{max}$ = 773 K
$p_{max}$ = 38 MPa
$\dot{m}_{max}$ = 12 kg h$^{-1}$
Pump(s)
Organic modifiers and/or metal salt solution
Pump(s)
Organic modifiers and/or metal salt solution
Pump(s)
Deionized water (filter backwashing)
Product dispersion
Back pressure regulator
Lift check valve
Filter unit 1
Filter unit 2
Waste
Cooler
Lift check valve
Lift check valve
$T_{salt}$
$p_{before mix}$
Mixing unit
$T_{hot water}$
Safety relief valve
$p_{after mix}$
Reactor
$T_{after mix}$
$T_{reactor}$
Cooler
$T_{cool}$
Waste
electrical heating blocks
Lift check valve
$F_{Quench}$
(Pulsation damper)
Pump(s)
Deionized water (quench)
Lift check valve
$F_{water}$
Syringe pumps
Deionized water (main water stream)

## Versuchsparamter und Messdaten

## –Nanoanlage–

20. Oktober 2008

| | |
|---|---|
| Versuchsname | **NAV5117092008c2** |
| Soll Druck [bar] | 300 |
| Soll Temperatur [°C] | 400 |
| Soll Dichte [kg/m$^3$] | 0,00 |
| Soll Volumenstrom [mL/min] | 100 |
| Soll Stromverhältnis | 0.2 |
| Salzart | FeAc/FeNO3 |
| Salzkonzentration (Edukt) [mol/L] | 0 |
| Salzkonzentration (Produkt) [mol/L] | 0 |
| pH (Produkt) | 2.98 |
| pH Anpassung | nein |
| Funktionalisierung | nein |
| Mischer | 4 |
| T Kreuz 1 [°C] | 400 |
| T Kreuz 2 [°C] | 406 |
| T Vorerhitzer [°C] | 433 |
| Volumenstrom Isco [mL/min] | 80 |
| Volumenstrom Isco Salz [mL/min] | 20 |
| Volumenstrom HPLC Salz [mL/min] | 0 |
| Volumenstrom HPLC Organik [mL/min] | 0 |
| Volumenstrom HPLC Salz/Organik [mL/min] | 0 |
| Hub Lewa 1 | 0 |
| Hub Lewa 2 | 357,00 |

| | Mittelwert | Std. Abw. | Min | Max | Spanne |
|---|---|---|---|---|---|
| Q Quench [kg/h] | 0.013 | 0.028 | -0.057 | 0.076 | 0.133 |
| Q Hauptstrom [kg/h] | 4.854 | 0.035 | 4.559 | 4.969 | 0.41 |
| P vor Reaktor [bar] | 303.41 | 0.29 | 302.47 | 304.09 | 1.62 |
| P nach Reaktor [bar] | 300.39 | 0.39 | 299.1 | 301.22 | 2.13 |
| T nach Vorerhitzer [°C] | 433.41 | 0.09 | 433.08 | 433.77 | 0.69 |
| T vor Reaktor [°C] | 400.22 | 0.12 | 399.75 | 400.78 | 1.02 |
| T Reaktor [°C] | 400.05 | 0.11 | 399.62 | 400.55 | 0.93 |
| T Organik-Salz-Strom [°C] | 21.5 | 0.37 | 20.86 | 22.16 | 1.3 |
| Pvor - Pnach [bar] | 3.02 | 0.29 | 2.08 | 3.86 | 1.78 |

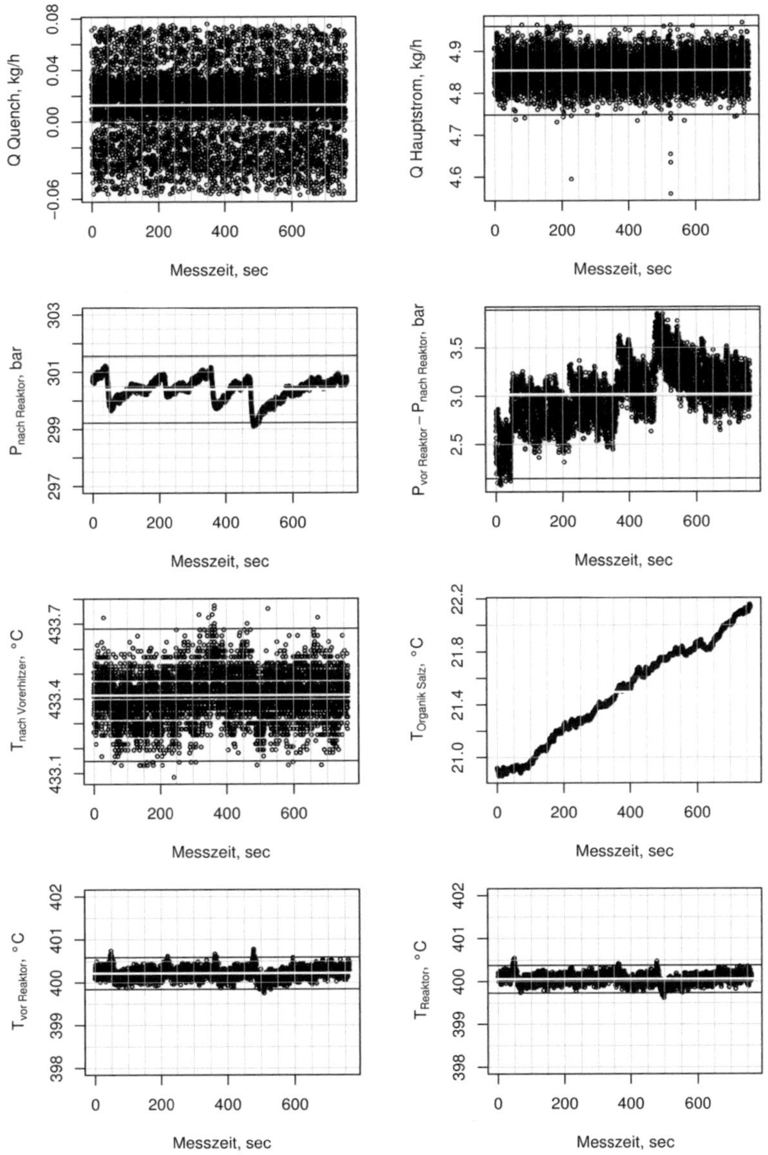


**Figure C.5:** Example file of the generated summary for each recorded experimental period.

## Check-Liste Nanoanlage

- Namen an Belegungstafel eintragen
- Warnhinweis muss sichtbar sein (Schutzbrille)
- Wasserbehälter füllen
- Wasser entgasen (1/2 bis ¾ Std.)
- Stromversorgung
    - 8 normale Stromstecker (evtl. 9 falls neuer Rechner auch laufen soll)
    - 2 Starkstromstecker (in Steckdose ohne FI-Sicherung!!!)
- Kühlwasser
    - 4 Hebel (Wasser muss zur Nanoanlage laufen)
    - Bewegt sich der Schwimmer?
    - Hand an den Kühler halten. Wird dieser Kühl? < 25°C
- Abzug anschalten
- Druckluft ein (für Pumpe)

- Computer
    - DasyLab starten
        - Verzeichnis: Datei->öffnen->Laufwerk C:/Nanoanlage/Schaltbilder/Messwerterfassung Nanoanlage mit Druckhandregler.DSB
        - Programm starten (grüner Pfeil, links oben)
          Kontrolle ob Sensoren funktionieren bzw. Programm läuft (Druck muss ein wenig schwanken)

- Ventile einstellen
    - Fließweg einstellen
    - Handregelventil einstellen
    - Zu Beginn sollte kein Druck aufgebaut werden können!!!

- Pumpen
    - Evtl. Kolben entlüften
    - Fließgeschwindigkeit einstellen

- Druckaufbau mit Druckhandregler
    - Prozessparameter einstellen

- Temperatur
  - Auf Temperatur einstellen, aber erst wenn der Druck (bsp. 300bar) erreicht ist.
  - !!! Vorerhitzer 3 muss von Hand geregelt werden. AN/AUS-Schalter hat keine Funktion, wenn Temperatur eingestellt wird, heizt das Element!!!!

**!!!Kühler kontrollieren, auch während des Versuchs!!!**

- Probenentnahme
  - Ist Zustand stationär? (Druck, Temperatur, pH-Wert...)
  - Wurde lang genug gewartet, dass auch wirklich das Produkt aus der Anlage kommt (siehe Anhang oder Buch Kapitel „Wartezeiten" Seite 77)
  - min. 7-10 Minuten warten, je nach Aufbau der Anlage
  - Schreiber einschalten, 2 Minuten warten, dann kann die Probe entnommen werden
  - Schreiber nach Probenentnahme ausschalten

Falls nötig, die Anlage zwischen den einzelnen Versuchen mit entgastem Wasser spülen und Druckstöße zur Beseitigung von Ablagerungen durchführen.

**!!! Druck muss zwischen 250 und 320 bar sein !!!**

- Versuchsende
  - Alle Thermostate ausschalten
    **!!!Vorerhitzer 3 muss von Hand auf 0 gestellt werden!!!**
  - **!!!Druck darf erst herausgenommen werden, wenn die Temperatur unter 100 °C ist.**
  - Eventuell nochmalige Druckstöße um Verunreinigungen zu lösen/entfernen

**Figure C.6:** Check list for the safe operation of the pilot plant.

# Appendix D

# Additional data for Chapter 5

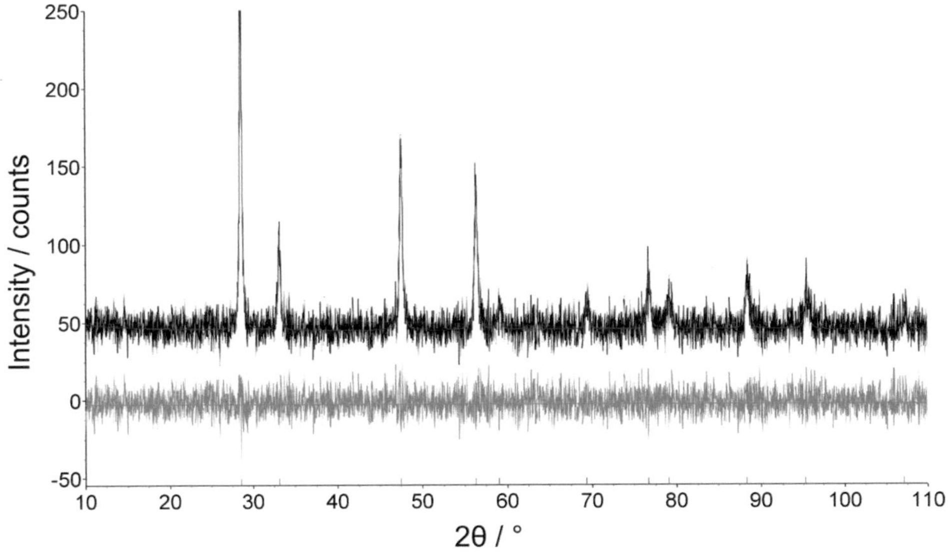


**Figure D.1:** PXRD whole pattern fit for sample V3: experimental data in black, fit in light gray and difference in dark gray.

# Appendix E

# Additional data for Chapter 6

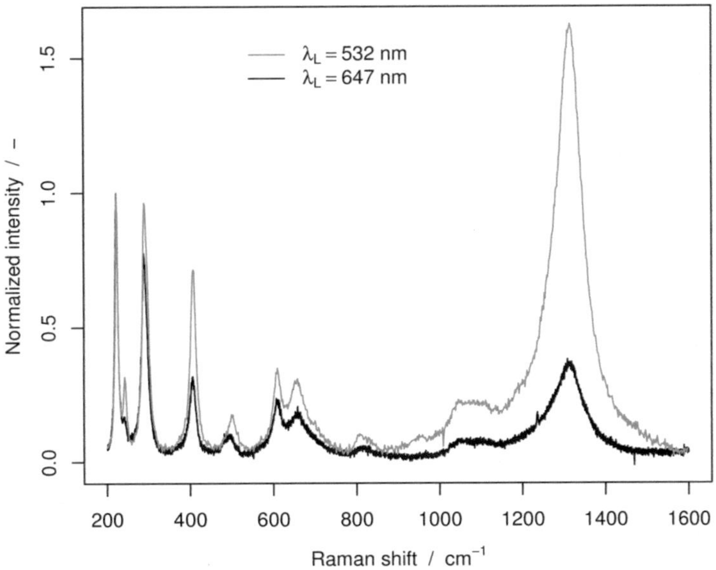


**Figure E.1:** Raman spectra of hematite nanoparticles prepared according to Schwertmann and Cornell (2000) taken using different lasers.

**Figure E.2** ***(facing page)*****:** Powder X-ray diffraction of the reference samples for the Raman spectroscopic investigations. The labeling is in accordance to Figure 6.3; positions and relative intensities of the lines for the relevant iron oxide phases are shown at the bottom (PDF 33-664 for $\alpha$-$Fe_2O_3$, PDF 19-629 for $Fe_3O_4$, PDF 39-1346 for $\gamma$-$Fe_2O_3$ from the International Centre for Diffcation Data, ICDD, USA); PXRD measurements of the particles were performed with a Bragg-Brentano-Goniometer D8 (Bruker AXS) using monochromatic Cu K$\alpha$ radiation, a secondary monochromator, sample changer and a scintillation counter.

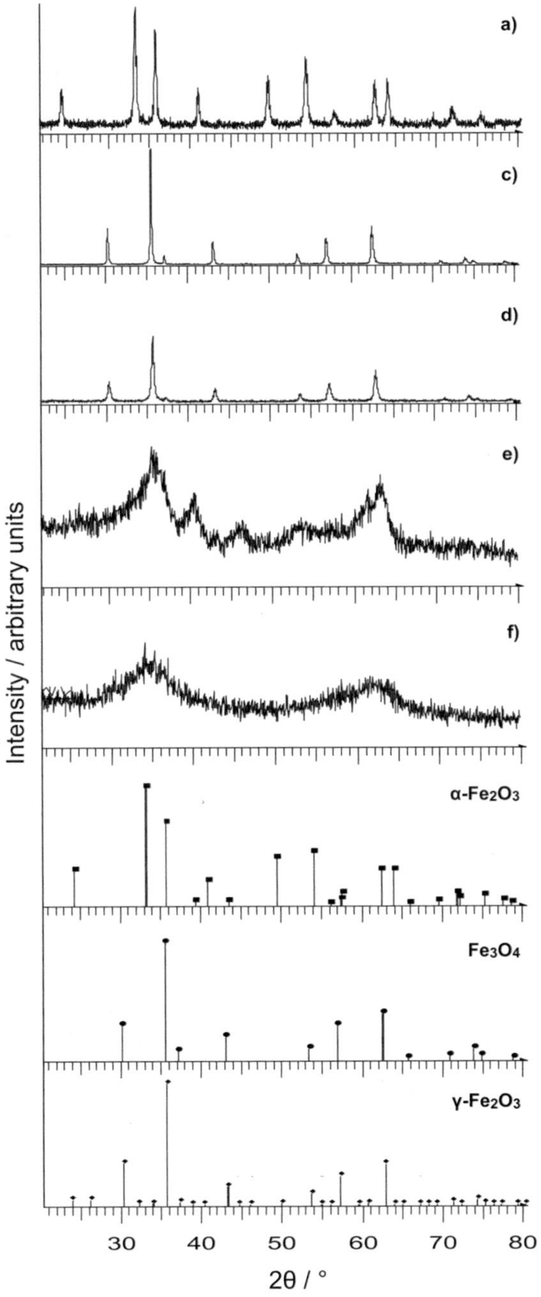
a)
c)
d)
e)
f)
α-Fe2O3
Fe3O4
γ-Fe2O3
Intensity / arbitrary units
30
40
50
60
70
80
2θ / °

**Table E.1:** Mean Crystallite sizes and composition of some of the reference samples used for the Raman spectroscopic investigations. The labeling is in accordance to Figure 6.3. The volume weighted crystallite sizes were determined from the Lorentz width of the diffraction peaks. Correction for the instrument broadening was performed using the standard reference material 660a ($LaB_6$) of the National Institutes of Standards & Technology (NIST). The crystalline phases present in the samples were identified by comparing with the positions and relative intensities of the lines of relevant iron oxide phases from the International Centre for Diffraction Data (ICDD, USA).

| | | **Crystallite size [nm]** | | **Composition [%]** | |
|---|---|---|---|---|---|
| **Label** | **Reference Sample** | $\alpha$-$Fe_2O_3$ | $\gamma$-$Fe_2O_3$ or $Fe_3O_4$ | $\alpha$-$Fe_2O_3$ | $\gamma$-$Fe_2O_3$ or $Fe_3O_4$ |
| a) | hematite nanoparticles Schwertmann and Cornell (2000) | 40 | - | 100 | - |
| c) | iron(II, III) oxide powder (<5 µm, 95%) | - | 93 | - | 100 |
| d) | iron(II,III) oxide nano-powder (<50 nm particle size (TEM) | - | 33 | - | 100 |

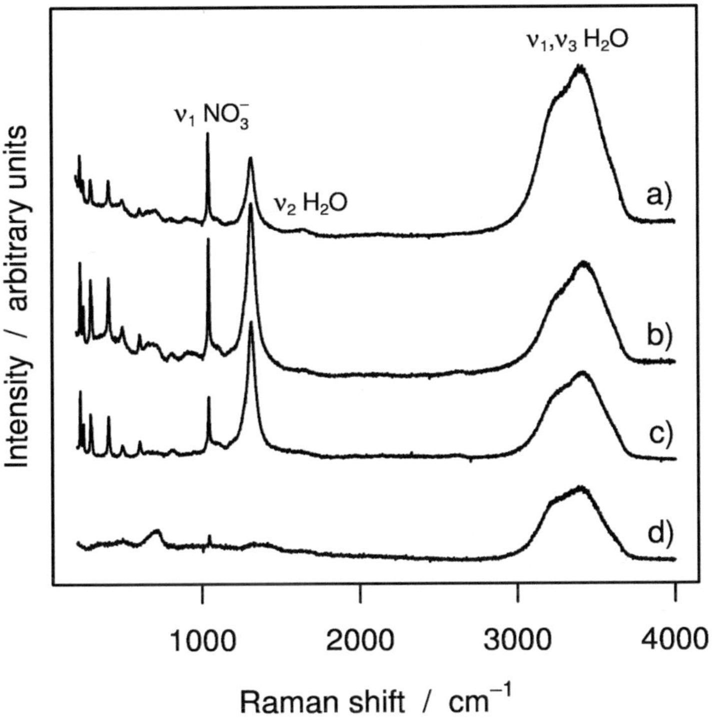


**Figure E.3:** Overview Raman spectra of iron oxide nanoparticle dispersions synthesized at $\phi_m = 0.2$ and $\dot{m}_{total} = 6.07$ kg/h: a) using iron(III) nitrate with $c_{\text{product Fe}} = 0.0055$ mol/L taken after increasing the iron oxide content via evaporation of water under vacuum up to about 30 g/L; b) same as a) but with an iron oxide content of about 85 g/L; c) using iron(III) nitrate with $c_{\text{product Fe}} = 0.001$ mol/L, measured as a wet precipitate; d) using iron(II) acetate with the addition of 0.002 mol/L $HNO_3$ and with $c_{\text{product Fe}} = 0.001$ mol/L, measured as a concentrated dispersion.

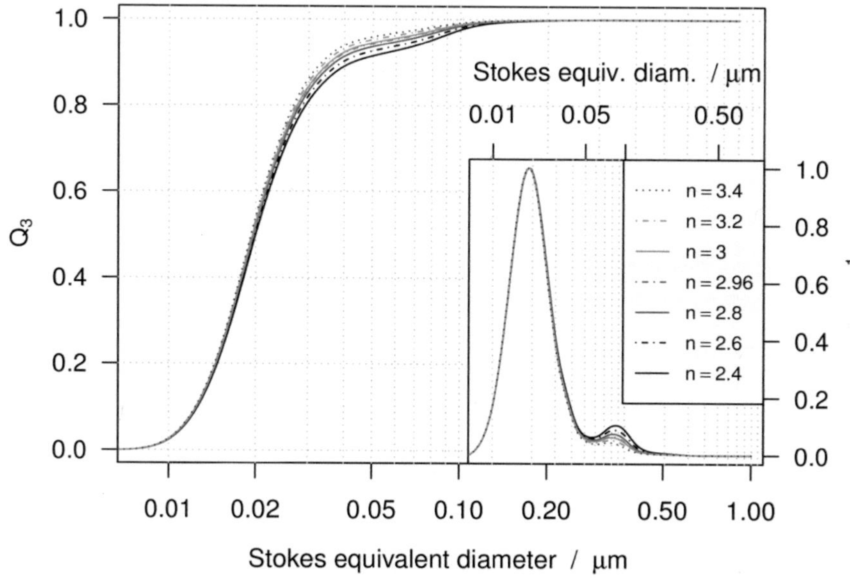


**Figure E.4:** Characteristic values of the particle size distributions synthesized from iron(III) nitrate varying the assumed real part of the refractive index.

**Table E.2:** Characteristic values of the particle size distributions synthesized from iron(III) nitrate varying the assumed real part of the refractive index. Values were chosen within the range of the reported values in Glasscock et al. (2008); Hellwege and Hellwege (1962); Longtin et al. (1988). The values used to compare the dispersions obtained under different conditions are displayed in gray.

| $n$ | $x_{k,3}$ nm | $x_{50,3}$ nm | $x_{10,3}$ nm | $x_{90,3}$ nm | $x_{g,3}$ nm | $\sigma_{g,3}$ |
|---|---|---|---|---|---|---|
| 2.4 | 18.8 | 20 | 12.6 | 42.7 | 22.08 | 1.72 |
| 2.6 | 18.7 | 19.8 | 12.6 | 39.3 | 21.61 | 1.68 |
| 2.8 | 18.7 | 19.6 | 12.5 | 37.1 | 21.21 | 1.64 |
| 2.96 | 18.7 | 19.5 | 12.5 | 35.8 | 20.92 | 1.62 |
| 3 | 18.7 | 19.5 | 12.5 | 35.5 | 20.86 | 1.61 |
| 3.2 | 18.7 | 19.4 | 12.4 | 34.4 | 20.55 | 1.58 |
| 3.4 | 18.7 | 19.3 | 12.4 | 33.4 | 20.28 | 1.56 |

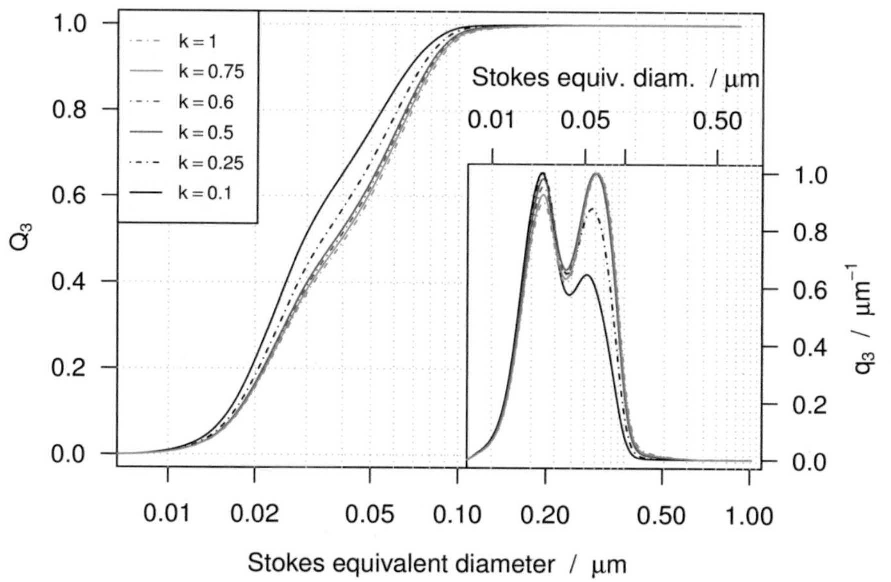


**Figure E.5:** Characteristic values of the particle size distributions synthesized from iron(III) nitrate varying the assumed absorption value ($k$).

**Table E.3:** Characteristic values of the particle size distributions synthesized from iron(III) nitrate varying the assumed absorption value. The absorption value was varied within the full range of values for strongly coloured metal oxides (Müller and Schuhmann, 1997). The highest and lowest absorption value reported in the considered literature is about 0.25 and 0.6 respectively (Glasscock et al., 2008; Longtin et al., 1988). The values used to compare the dispersions obtained under different conditions are displayed in gray.

| $k$ | $x_{k,3}$ | $x_{50,3}$ | $x_{10,3}$ | $x_{90,3}$ | $x_{g,3}$ | $\sigma_{g,3}$ |
|---|---|---|---|---|---|---|
| - | nm | nm | nm | nm | nm | |
| 0.1 | 18.6 | 19 | 12.3 | 31.2 | 19.6 | 1.49 |
| 0.25 | 18.7 | 19.4 | 12.4 | 33.7 | 20.35 | 1.56 |
| 0.4 | 18.7 | 19.4 | 12.5 | 35 | 20.73 | 1.6 |
| 0.5 | 18.7 | 19.5 | 12.5 | 35.8 | 20.92 | 1.62 |
| 0.6 | 18.7 | 19.6 | 12.5 | 36.5 | 21.08 | 1.63 |
| 0.75 | 18.7 | 19.6 | 12.5 | 37.4 | 21.28 | 1.66 |
| 1 | 18.7 | 19.7 | 12.5 | 38.7 | 21.54 | 1.69 |

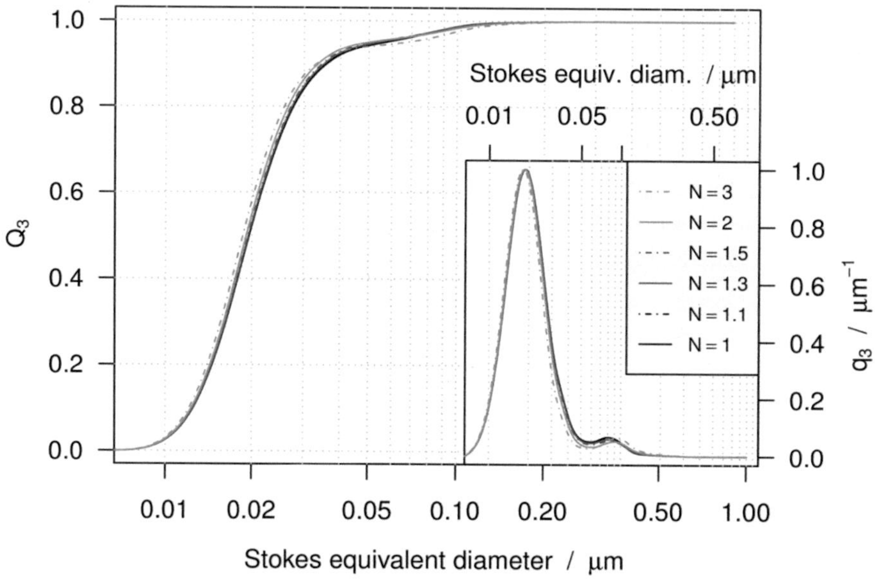


**Figure E.6:** Characteristic values of the particle size distributions synthesized from iron(III) nitrate varying the non-sphericity factor (*s*).

**Table E.4:** Characteristic values of the particle size distributions synthesized from iron(III) nitrate varying the assumed non-sphericity factor ($s$); this is the average aspect ratio (longest to shortest dimension) for all possible orientations of the particles. The values used to compare the dispersions obtained under different conditions are displayed in gray.

| $s$ | $x_{k,3}$ nm | $x_{50,3}$ nm | $x_{10,3}$ nm | $x_{90,3}$ nm | $x_{g,3}$ nm | $\sigma_{g,3}$ |
|---|---|---|---|---|---|---|
| 1 | 18.7 | 19.5 | 12.5 | 36 | 20.96 | 1.62 |
| 1.1 | 18.7 | 19.5 | 12.5 | 35.8 | 20.92 | 1.62 |
| 1.3 | 18.7 | 19.4 | 12.5 | 35.2 | 20.8 | 1.61 |
| 1.5 | 18.7 | 19.4 | 12.4 | 34.7 | 20.68 | 1.61 |
| 2 | 18.6 | 19.2 | 12.3 | 33.6 | 20.42 | 1.61 |
| 3 | 18 | 18.6 | 12.1 | 33.2 | 20.2 | 1.67 |

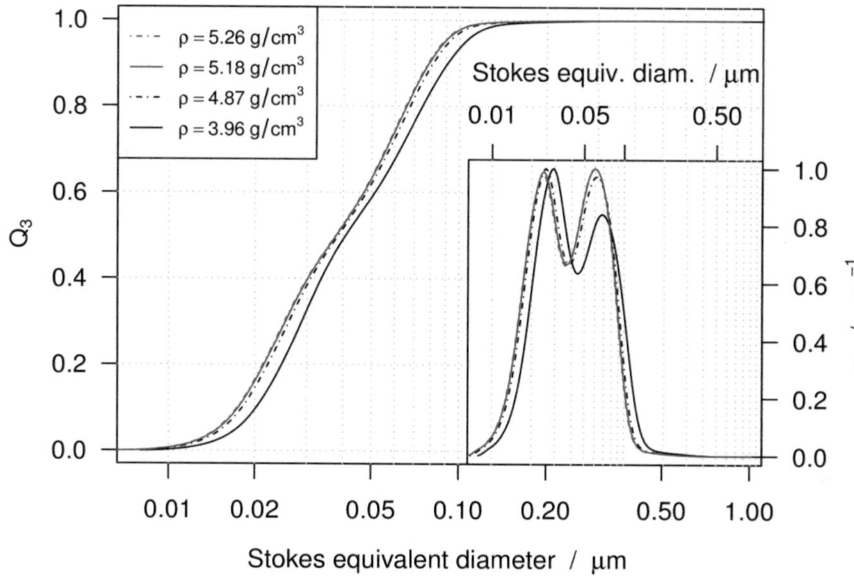


**Figure E.7:** Characteristic values of the particle size distributions synthesized from iron(III) nitrate with a high fraction of bigger particles varying the assumed density.

**Table E.5:** Characteristic values of the particle size distributions synthesized from iron(III) nitrate with a high fraction of bigger particles varying the assumed density; The densities of ferrihydrite ($\rho = 3.96$ g/cm$^3$) and the bulk densities of relevant iron oxides (in the order of maghemite, magnetite and hematite) were taken from Cornell and Schwertmann (2003). The values used to compare the dispersions obtained under different conditions are displayed in gray.

| $\rho$ g/cm$^3$ | $x_{k,3}$ nm | $x_{50,3}$ nm | $x_{10,3}$ nm | $x_{90,3}$ nm | $x_{g,3}$ nm | $\sigma_{g,3}$ |
|---|---|---|---|---|---|---|
| 3.96 | 29.1 | 40.8 | 20.2 | 91.2 | 42.38 | 1.79 |
| 4.87 | 25.7 | 38.5 | 18.1 | 81.2 | 38.47 | 1.79 |
| 5.18 | 60 | 37.9 | 17.5 | 78.6 | 37.43 | 1.79 |
| 5.26 | 59.5 | 37.7 | 17.4 | 78 | 37.18 | 1.79 |

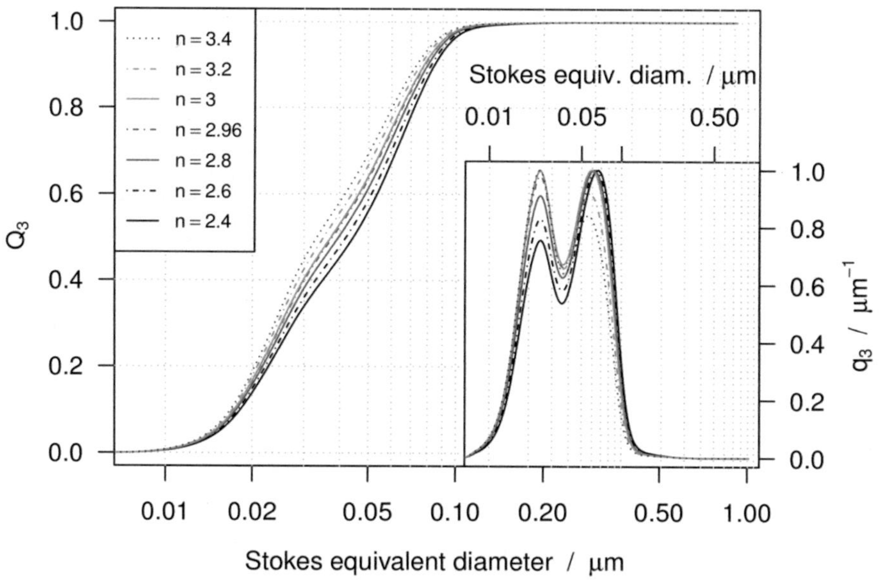


**Figure E.8:** Characteristic values of the particle size distributions synthesized from iron(III) nitrate with a high fraction of bigger particles varying the assumed real part of the refractive index.

**Table E.6:** Characteristic values of the particle size distributions synthesized from iron(III) nitrate with a high fraction of bigger particles varying the assumed real part of the refractive index. Values were chosen within the range of the reported values in Glasscock et al. (2008); Hellwege and Hellwege (1962); Longtin et al. (1988). The values used to compare the dispersions obtained under different conditions are displayed in gray.

| $n$ | $x_{k,3}$ nm | $x_{50,3}$ nm | $x_{10,3}$ nm | $x_{90,3}$ nm | $x_{g,3}$ nm | $\sigma_{g,3}$ |
|---|---|---|---|---|---|---|
| 2.4 | 66.8 | 44.6 | 18.2 | 84.3 | 41.03 | 1.82 |
| 2.6 | 62.9 | 42.1 | 17.9 | 82.1 | 39.6 | 1.81 |
| 2.8 | 62.8 | 39.6 | 17.5 | 79.8 | 38.23 | 1.8 |
| 2.96 | 59.5 | 37.7 | 17.4 | 78 | 37.18 | 1.79 |
| 3 | 59.5 | 37.3 | 17.3 | 77.5 | 36.92 | 1.79 |
| 3.2 | 24.4 | 35.1 | 17.1 | 75.3 | 35.68 | 1.78 |
| 3.4 | 24.2 | 33.2 | 16.8 | 73 | 34.5 | 1.77 |

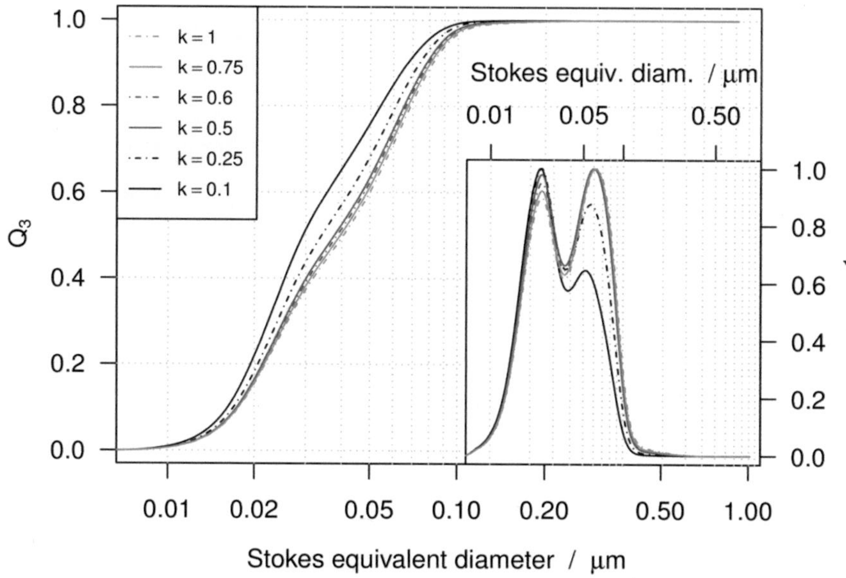


**Figure E.9:** Characteristic values of the particle size distributions synthesized from iron(III) nitrate with a high fraction of bigger particles varying the assumed absorption value ($k$).

**Table E.7:** Characteristic values of the particle size distributions synthesized from iron(III) nitrate with a high fraction of bigger particles varying the absorption value ($k$). The values used to compare the dispersions obtained under different conditions are displayed in gray.

| $k$ | $x_{k,3}$ | $x_{50,3}$ | $x_{10,3}$ | $x_{90,3}$ | $x_{g,3}$ | $\sigma_{g,3}$ |
|---|---|---|---|---|---|---|
| - | nm | nm | nm | nm | nm | |
| 0.1 | 23.9 | 29.4 | 16.1 | 67.3 | 31.55 | 1.72 |
| 0.25 | 24.4 | 34 | 16.9 | 73.9 | 34.96 | 1.76 |
| 0.4 | 24.4 | 36.6 | 17.3 | 76.7 | 36.51 | 1.78 |
| 0.5 | 59.5 | 37.7 | 17.4 | 78 | 37.18 | 1.79 |
| 0.6 | 60.5 | 38.6 | 17.5 | 79.1 | 37.7 | 1.8 |
| 0.75 | 62.8 | 39.5 | 17.5 | 80.3 | 38.29 | 1.81 |
| 1 | 62.8 | 40.6 | 17.6 | 82 | 38.99 | 1.82 |

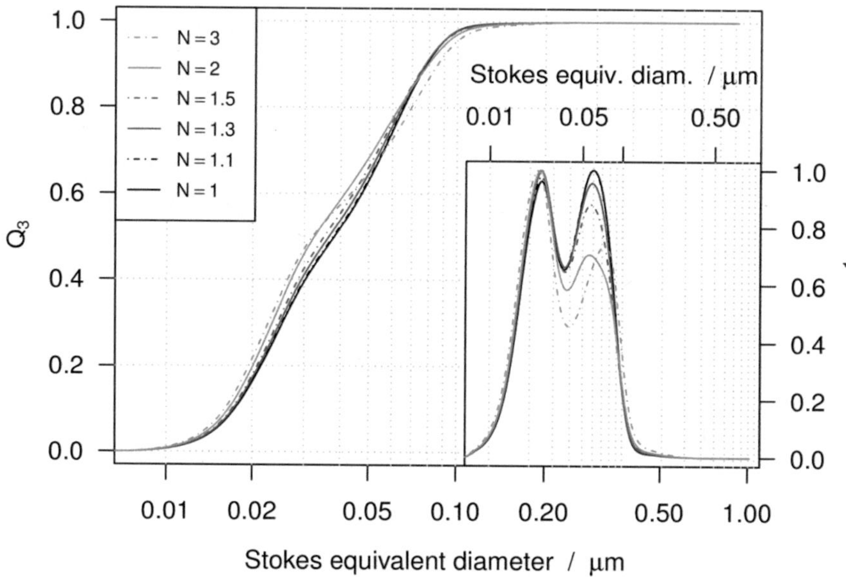


**Figure E.10:** Characteristic values of the particle size distributions synthesized from iron(III) nitrate with a high fraction of bigger particles varying the non-sphericity factor (*s*); this is the average aspect ratio (longest to shortest dimension) for all possible orientations of the particles.

**Table E.8:** Characteristic values of the particle size distributions synthesized from iron(III) nitrate with a high fraction of bigger particles varying the assumed non-sphericity factor ($s$); this is the average aspect ratio (longest to shortest dimension) for all possible orientations of the particles. The values used to compare the dispersions obtained under different conditions are displayed in gray.

| $s$ | $x_{k,3}$ nm | $x_{50,3}$ nm | $x_{10,3}$ nm | $x_{90,3}$ nm | $x_{g,3}$ nm | $\sigma_{g,3}$ |
|---|---|---|---|---|---|---|
| 1 | 60.3 | 38.1 | 17.4 | 78.2 | 37.36 | 1.79 |
| 1.1 | 59.5 | 37.7 | 17.4 | 78 | 37.18 | 1.79 |
| 1.3 | 24.3 | 36.5 | 17.2 | 77.5 | 36.58 | 1.79 |
| 1.5 | 24.3 | 35.1 | 17 | 77.3 | 35.94 | 1.8 |
| 2 | 23.9 | 32.6 | 16.5 | 79.2 | 34.88 | 1.83 |
| 3 | 22.6 | 32.1 | 16 | 85.4 | 35.58 | 1.92 |

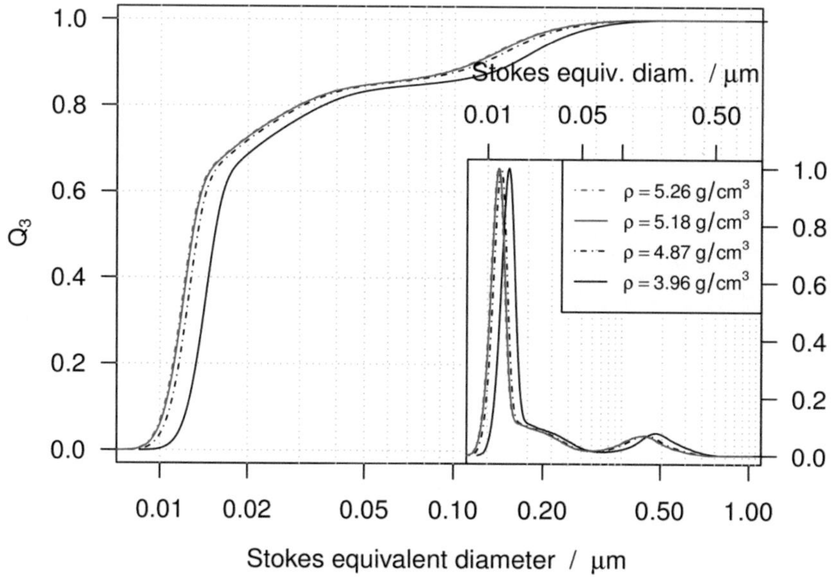


**Figure E.11:** Characteristic values of the particle size distributions synthesized from a mixture of iron(III) nitrate and iron(II) acetate varying the assumed density.

**Table E.9:** Characteristic values of the particle size distributions synthesized from a mixture of iron(III) nitrate and iron(II) acetate varying the assumed density; the densities of ferrihydrite ($\rho = 3.96$ g/cm$^3$) and the bulk densities of relevant iron oxides (in the order of maghemite, magnetite and hematite taken from Cornell and Schwertmann (2003). The values used to compare the dispersions obtained under different conditions are displayed in gray.

| $\rho$ g/cm$^3$ | $x_{k,3}$ nm | $x_{50,3}$ nm | $x_{10,3}$ nm | $x_{90,3}$ nm | $x_{g,3}$ nm | $\sigma_{g,3}$ |
|---|---|---|---|---|---|---|
| 3.96 | 14.4 | 15.4 | 12.5 | 157.2 | 23.84 | 2.54 |
| 4.87 | 12.6 | 13.4 | 10.9 | 123.6 | 20.42 | 2.46 |
| 5.18 | 12.1 | 12.9 | 10.5 | 116.2 | 19.61 | 2.45 |
| 5.26 | 12 | 12.8 | 10.4 | 114.5 | 19.42 | 2.44 |

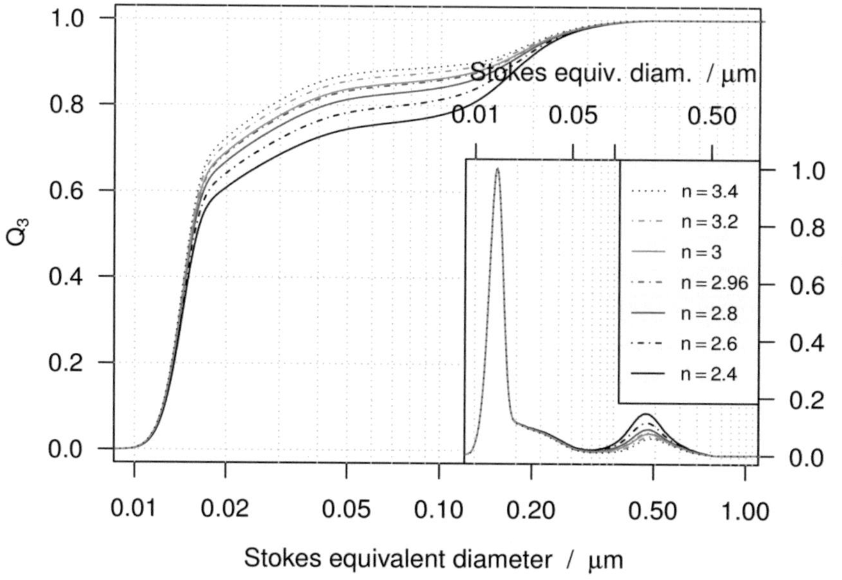


**Figure E.12:** Characteristic values of the particle size distributions synthesized from a mixture of iron(III) nitrate and iron(II) acetate varying the assumed real part of the refractive index.

**Table E.10:** Characteristic values of the particle size distributions synthesized from a mixture of iron(III) nitrate and iron(II) acetate varying the assumed real part of the refractive index. Values were chosen within the range of the reported values for hematite in Glasscock et al. (2008); Hellwege and Hellwege (1962); Longtin et al. (1988). The values used to compare the dispersions obtained under different conditions are displayed in gray. The assumed complex refractive index was chosen to be the same as for the dispersions synthesized using iron(III)nitrate which is most likely too high. However, except for the fraction of the bigger particles, the shape of the particle size distribution is only marginally affected by these values because the size distributions are very narrow. Due to the high uncertainty of the values characterizing the right part of the size distribution when larger particles are present in the investigated dispersions, these are put in parentheses when given in the main text.

| $n$ | $x_{k,3}$ nm | $x_{50,3}$ nm | $x_{10,3}$ nm | $x_{90,3}$ nm | $x_{g,3}$ nm | $\sigma_{g,3}$ |
|---|---|---|---|---|---|---|
| 2.4 | 14.4 | 16 | 12.6 | 184.6 | 29.29 | 2.89 |
| 2.6 | 14.4 | 15.7 | 12.5 | 175.2 | 26.78 | 2.75 |
| 2.8 | 14.4 | 15.5 | 12.5 | 165.2 | 24.95 | 2.62 |
| 2.96 | 14.4 | 15.4 | 12.5 | 157.2 | 23.84 | 2.54 |
| 3 | 14.4 | 15.4 | 12.5 | 155 | 23.6 | 2.52 |
| 3.2 | 14.4 | 15.2 | 12.3 | 143.4 | 22.59 | 2.43 |
| 3.4 | 14.4 | 15.1 | 12.3 | 127.8 | 21.8 | 2.36 |

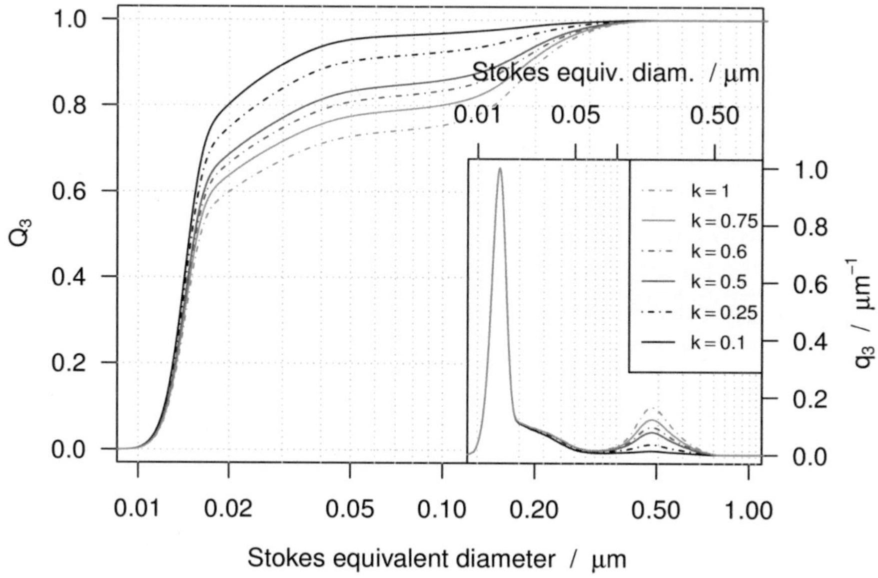


**Figure E.13:** Characteristic values of the particle size distributions synthesized from a mixture of iron(III) nitrate and iron(II) acetate varying the assumed absorption value ($k$).

**Table E.11:** Characteristic values of the particle size distributions synthesized from a mixture of iron(III) nitrate and iron(II) acetate varying the absorption value ($k$). The values used to compare the dispersions obtained under different conditions are displayed in gray. The assumed complex refractive index was chosen to be the same as for the dispersions synthesized using iron(III)nitrate which is most likely too high. However, except for the fraction of the bigger particles, the shape of the particle size distribution is only marginally affected by these values because the size distributions are very narrow. Due to the high uncertainty of the values characterizing the right part of the size distribution when larger particles are present in the investigated dispersions, these are put in parentheses when given in the main text.

| $k$ | $x_{k,3}$ | $x_{50,3}$ | $x_{10,3}$ | $x_{90,3}$ | $x_{g,3}$ | $\sigma_{g,3}$ |
|---|---|---|---|---|---|---|
| - | nm | nm | nm | nm | nm | |
| 0.1 | 14.4 | 14.9 | 12.2 | 30.6 | 17.6 | 1.73 |
| 0.25 | 14.4 | 15 | 12.3 | 47.6 | 20.02 | 2.1 |
| 0.4 | 14.4 | 15.2 | 12.3 | 130.3 | 22.31 | 2.38 |
| 0.5 | 14.4 | 15.4 | 12.5 | 157.2 | 23.84 | 2.54 |
| 0.6 | 14.4 | 15.5 | 12.5 | 172.2 | 25.35 | 2.67 |
| 0.75 | 14.4 | 15.7 | 12.5 | 187.4 | 27.56 | 2.85 |
| 1 | 14.4 | 16.1 | 12.6 | 203.7 | 30.97 | 3.06 |

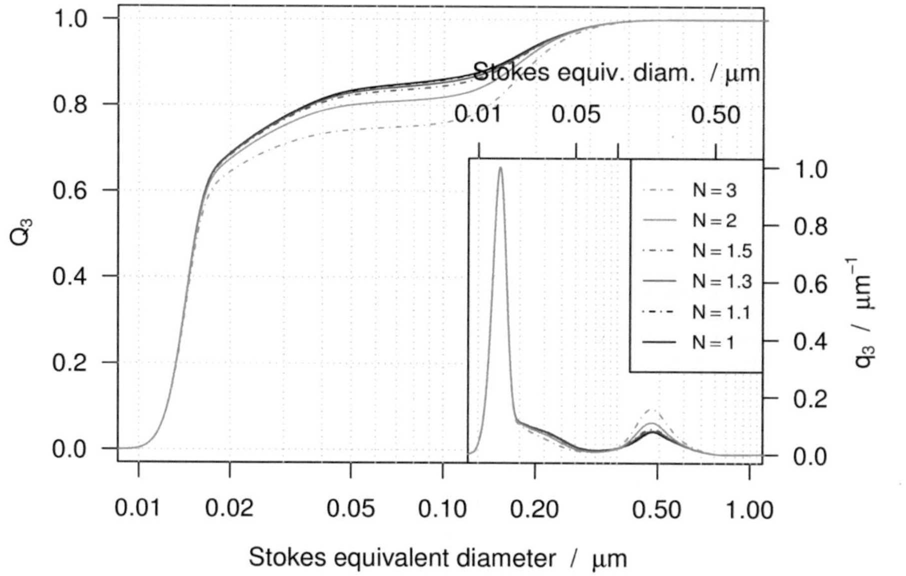


**Figure E.14:** Characteristic values of the particle size distributions synthesized from a mixture of iron(III) nitrate and iron(II) acetate varying the non-sphericity factor ($s$); this is the average aspect ratio (longest to shortest dimenstion) for all possible orientations of the particles.

**Table E.12:** Characteristic values of the particle size distributions synthesized from a mixture of iron(III) nitrate and iron(II) acetate varying the non-sphericity factor ($s$). The values used to compare the dispersions obtained under different conditions are displayed in gray.

| $s$ | $x_{k,3}$ nm | $x_{50,3}$ nm | $x_{10,3}$ nm | $x_{90,3}$ nm | $x_{g,3}$ nm | $\sigma_{g,3}$ |
|---|---|---|---|---|---|---|
| 1 | 14.4 | 15.4 | 12.5 | 156.1 | 23.78 | 2.53 |
| 1.1 | 14.4 | 15.4 | 12.5 | 157.2 | 23.84 | 2.54 |
| 1.3 | 14.4 | 15.4 | 12.5 | 160.7 | 24.06 | 2.57 |
| 1.5 | 14.4 | 15.4 | 12.5 | 164.5 | 24.39 | 2.61 |
| 2 | 14.4 | 15.5 | 12.5 | 176 | 25.61 | 2.73 |
| 3 | 14.4 | 15.6 | 12.5 | 195.6 | 29.11 | 3.02 |

**Figure E.15 *(facing page)*:** Overlays of ion-exchange chromatogramms: a) from the aqueous iron(II) acetate solution from the first packaging unit used ("first unit") and from a characteristic "normal" packaging unit ("different unit"). The first packing unit was used for the dispersions discussed in Section 3.2.2. (Dispersions synthesized from acidified aqueous iron acetate solutions using nitric acid) of the main text. b) from a reference solution containing 1000 ppm acetate and 10 ppm formate respectively; c) from the aqueous iron(II) acetate solution from the first packaging unit used ("first unit") and from a reference solution containing 10 ppm formate. The ion exchange chromatography was performed with an Alltech Anion exclusion column (length: 30 cm, i.d.: 7.8 mm) and a Alltech Wescan 315 conductivity detector using 2 mM sulfuric acid as eluent.

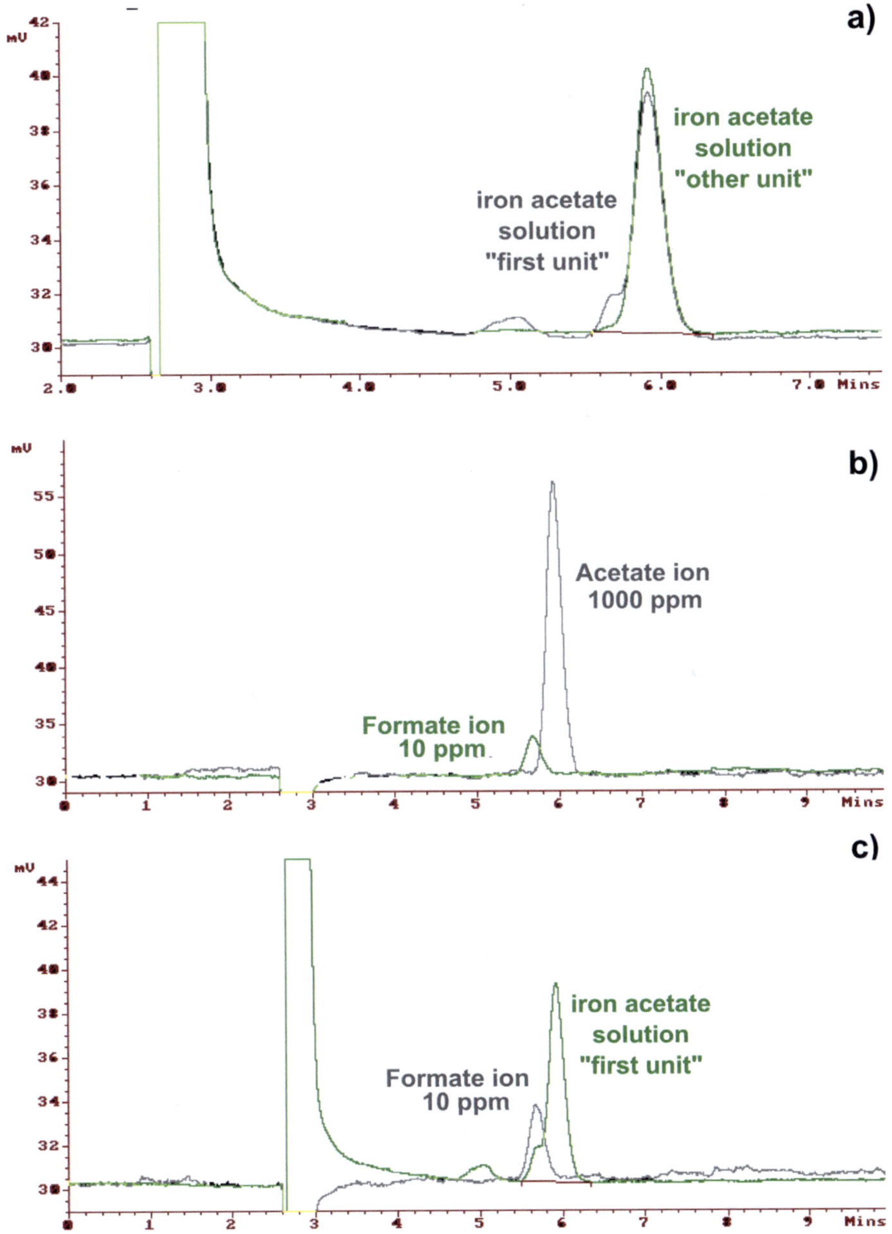
a)
mV
iron acetate
solution
"other unit"
iron acetate
solution
"first unit"
Mins
b)
mV
Acetate ion
1000 ppm
Formate ion
10 ppm
Mins
c)
mV
iron acetate
solution
"first unit"
Formate ion
10 ppm
Mins

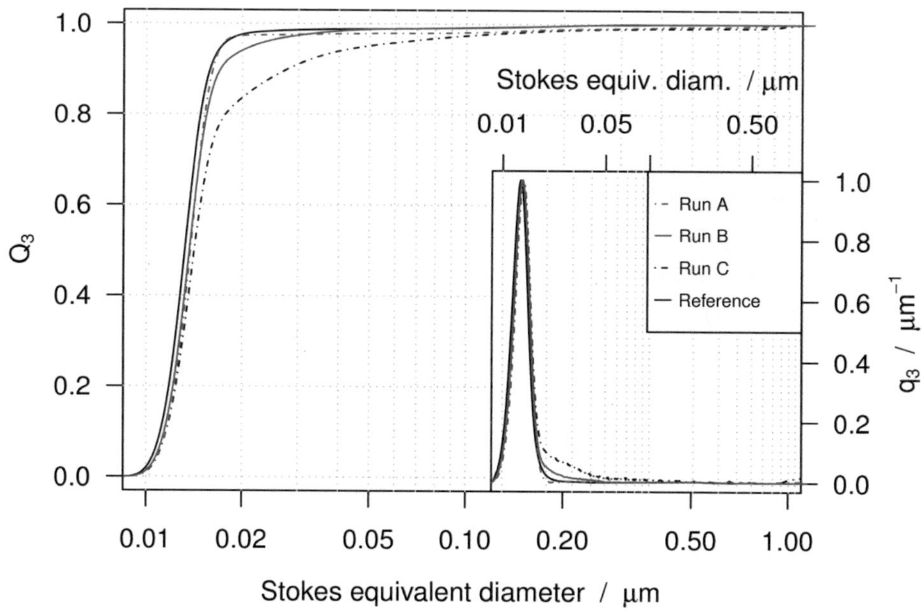


**Figure E.16:** Particle size distributions synthesized using a mixture of iron(II) acetate (65%) and iron(III) nitrate (35%) with the addition of 0.003 mol/L of acetic acid (acetate factor = 1.9) at $\dot{m}_{total}$ = 6.07 kg/h, $\phi_m$ = 0.2, and $c_{\text{product Fe}}$ = 0.001 mol/L, in comparison to the Reference sample (cf. Figure 6.13).

**Table E.13:** pH-values of the starting salt solution ($pH_{salt}$ / product ($pH_{product}$) from the dispersions synthesized from different compositions using mixtures of iron(II) acetate and iron(III) nitrate at $\dot{m}_{total}$ = 6.07 kg/h, $\phi_m$ = 0.2, and $c_{\text{product Fe}}$ = 0.001 mol/L.

| | | Addition of $HNO_3$ | | |
|---|---|---|---|---|
| Acetate factor | Fe(II):Fe(III) | 0 mmol/L | 2 mmol/L | 4 mmol/L |
| 1.7 | 75:25 | 4.5 / 3.3 | 3.9 / 3.1 | 3.2 / 3.0 |
| 1.7 | 65:35 | 3.9 / 3.4 | 3.1 / 3.0 | 2.9 / 2.9 |
| 1.7 | 55:45 | 2.9 / 2.9 | 2.9 / 2.8 | 2.9 / 2.9 |
| 1.9 | 75:25 | 4.2 / 3.2 | - / 3.1 | 3.1 / 2.9 |
| 1.9 | 65:35 | 3.7 / 3.1 | 2.9 / 2.9 | 2.6 / 2.8 |
| 1.9 | 55:45 | 3.1 / 2.9 | 2.9 / 2.8 | 2.7 / 2.7 |
| 2.1 | 55:45 | 3.1 / 2.9 | 2.8 / 2.7 | 2.4 / 2.7 |
| 2.1 | 45:55 | 2.8 / 2.8 | 2.8 / 2.7 | 2.5 / 2.6 |

**Table E.14:** Geometric standard deviation ($\sigma_{g,3}$, $\pm$0.09) of the particle size distribution (obtained by differential sedimentation analysis) from the dispersions synthesized from different compositions using mixtures of iron(II) acetate and iron(III) nitrate at $\dot{m}_{total}$ = 6.07 kg/h, $\phi_m$ = 0.2, and $c_{\text{product Fe}}$ = 0.001 mol/L.

| | | Addition of $HNO_3$ | | |
|---|---|---|---|---|
| Acetate factor | Fe(II):Fe(III) | 0 mmol/L | 2 mmol/L | 4 mmol/L |
| 1.7 | 75:25 | 1.46 | 1.65 | 1.58 |
| 1.7 | 65:35 | 1.63 | 1.38 | 1.50 |
| 1.7 | 55:45 | 1.32 | 1.35 | 1.54 |
| 1.9 | 75:25 | 1.58 | 1.65 | 1.80 |
| 1.9 | 65:35 | 1.37 | 1.45 | 2.06 |
| 1.9 | 55:45 | 1.45 | 1.37 | 1.36 |
| 2.1 | 55:45 | 1.54 | 1.54 | 1.47 |
| 2.1 | 45:55 | 1.58 | 1.46 | 1.51 |

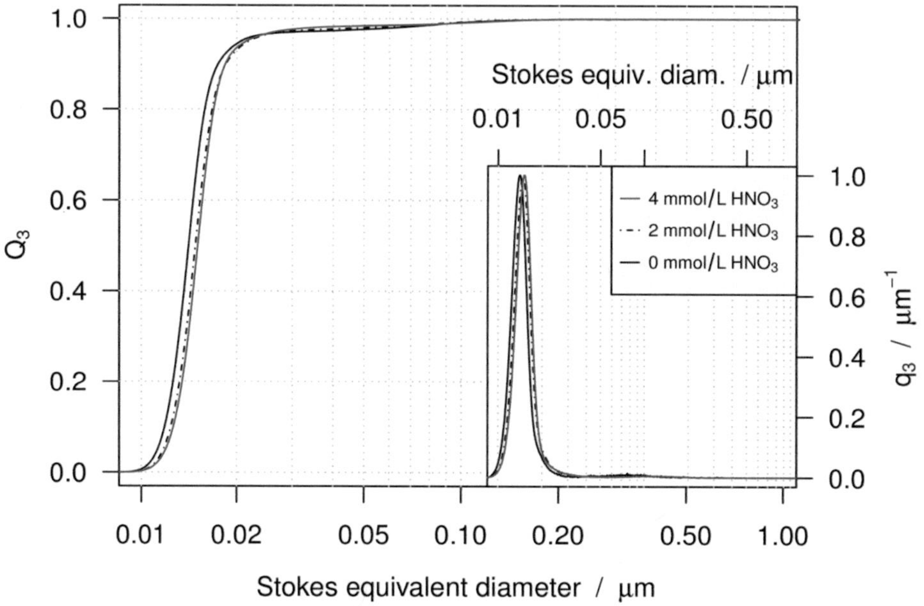


**Figure E.17:** Particle size distribution of iron oxide nanoparticle dispersions synthesized with $\phi_m$ = 0.2 and $\dot{m}_{total}$ = 6.07 kg/h using a mixture of iron(II) acetate (55%) and iron(III) nitrate (45%) with an acetate factor of 1.9 and $c_{\text{product Fe}}$ = 0.001 mol/L for different amounts of added nitric acid.

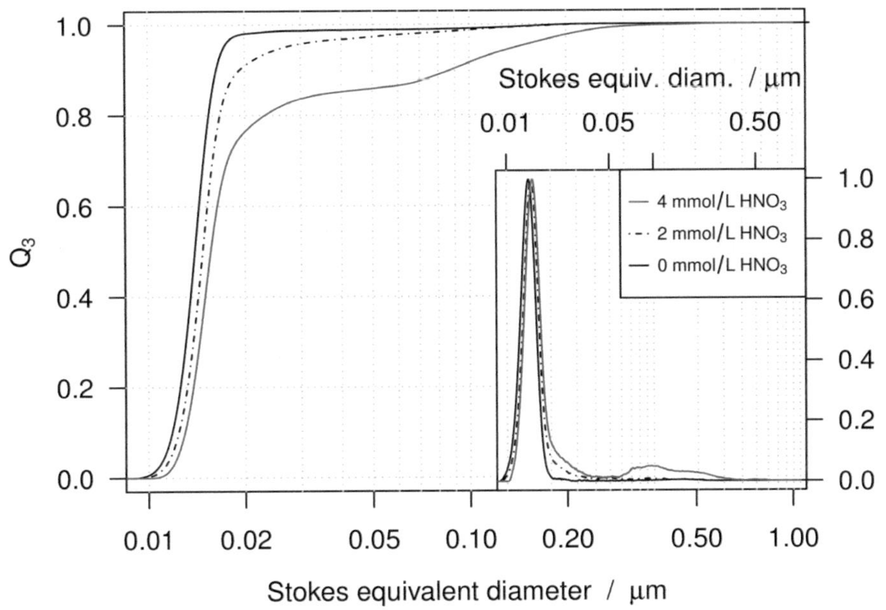


**Figure E.18:** Particle size distribution of iron oxide nanoparticle dispersions synthesized with $\phi_m$ = 0.2 and $\dot{m}_{total}$ = 6.07 kg/h using a mixture of iron(II) acetate (65%) and iron(III) nitrate (35%) with an acetate factor of 1.9 and $c_{\text{product Fe}}$ = 0.001 mol/L for different amounts of added nitric acid.

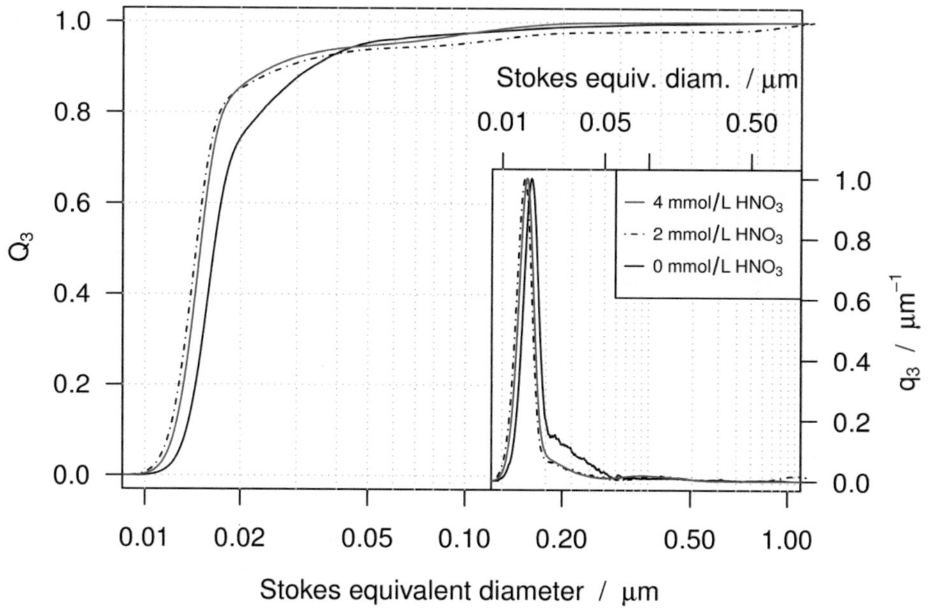


**Figure E.19:** Particle size distribution of iron oxide nanoparticle dispersions synthesized with $\phi_m = 0.2$ and $\dot{m}_{total} = 6.07$ kg/h using a mixture of iron(II) acetate (75%) and iron(III) nitrate (25%) with an acetate factor of 1.9 and $c_{\text{product Fe}} = 0.001$ mol/L for different amounts of added nitric acid.

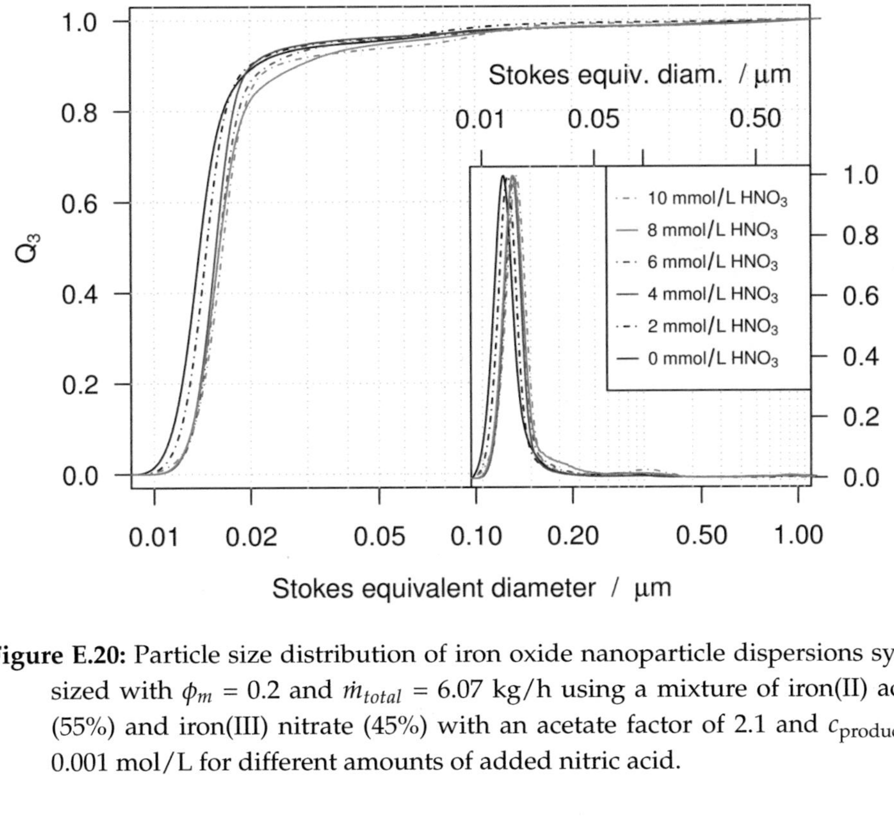


**Figure E.20:** Particle size distribution of iron oxide nanoparticle dispersions synthesized with $\phi_m = 0.2$ and $\dot{m}_{total} = 6.07$ kg/h using a mixture of iron(II) acetate (55%) and iron(III) nitrate (45%) with an acetate factor of 2.1 and $c_{\text{product Fe}} = 0.001$ mol/L for different amounts of added nitric acid.

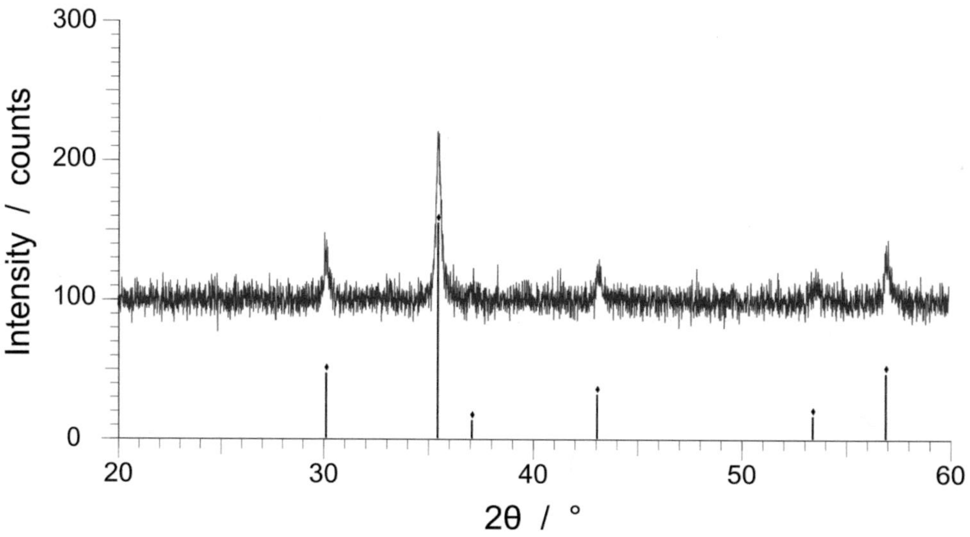


**Figure E.21:** Powder X-ray diffraction pattern corresponding to the particles in Figure 6.10. Vertical lines mark the position and represent the (relative) intensity for magnetite (PDF 19-629 from the International Centre for Diffraction Data, ICDD, USA).

# Appendix F

# Additional data for Chapter 7

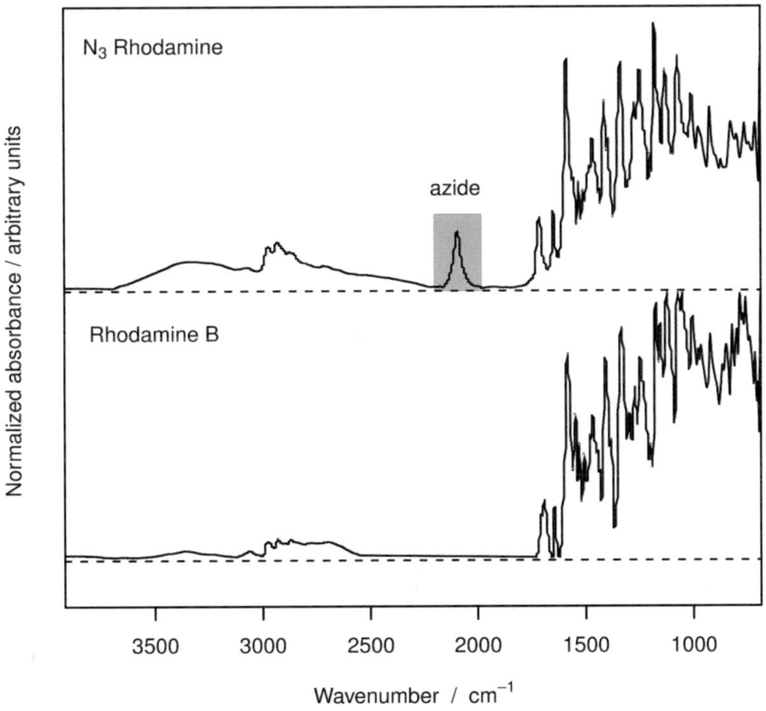


**Figure F.1:** Infrared spectra indicating the successful linkage of the azide group to Rhodamine B.

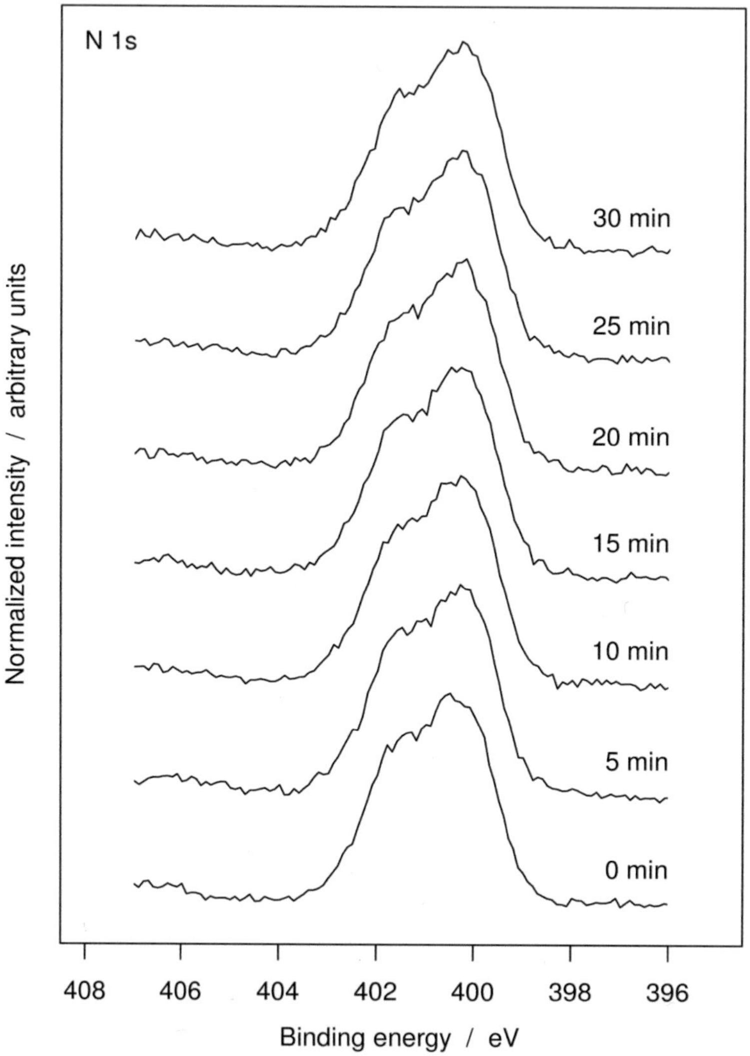


**Figure F.2:** XPS analysis of 4-(1-benzyl-1H[1,2,3]triazol-4-yl)butyric acid (reference sample for "1,2,3-triazole ring" compounds); N 1s spectra observed after different exposure times.

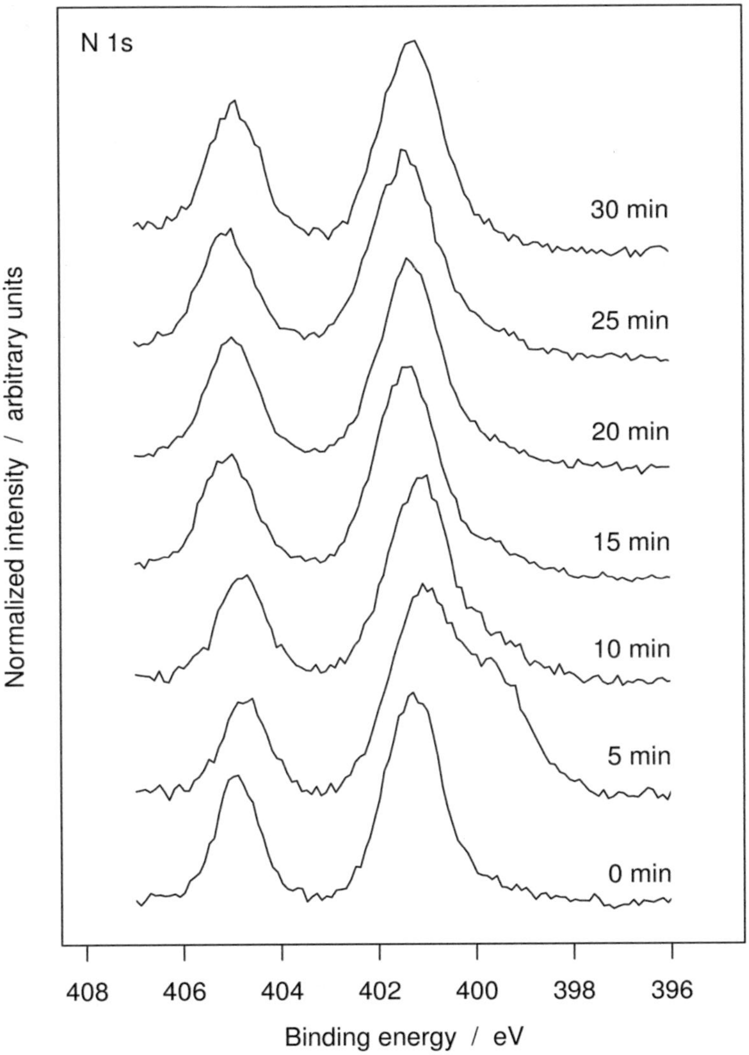


**Figure F.3:** XPS analysis of the galactose with azide-terminated linker (cf. Figure 7.16 on page 168 for structure); N 1s spectra observed after different exposure times.

# Publications

Many ideas, tables and figures contained in this thesis also appear in the following publications:

- M. Daschner de Tercero, C. Röder, U. Fehrenbacher, U. Teipel, M. Türk
  *Continuous supercritical hydrothermal synthesis of iron oxide nanoparticle dispersions and their characterization*
  Journal of Nanoparticle Research, 16:2350, 1-27, (2014)
- M. Daschner de Tercero, I. González Martínez, M. Herrmann, M. Bruns, C. Kübel, S. Jennewein, U. Fehrenbacher, L. Barner, M. Türk
  *Synthesis of in situ functionalized iron oxide nanoparticles presenting alkyne groups via a continuous process using near-critical and supercritical water*
  The Journal of Supercritical Fluids, 82, 83 - 95, (2013)
- M. Daschner de Tercero, M. Bruns, I. González Martínez, M. Türk, U. Fehrenbacher, S. Jennewein, L. Barner
  *Continuous Hydrothermal Synthesis of In Situ Functionalized Iron Oxide Nanoparticles: A General Strategy to Produce Metal Oxide Nanoparticles With Clickable Anchors*
  Particle & Particle Systems Characterization, 30, 229-23, (2013)
- M. Daschner de Tercero, S. Jennewein, U. Fehrenbacher, L. Barner
  *Hydrothermale Herstellung von Nanopartikeln mit clickfähigen Linkern an der Oberfläche*
  Deutsche Patentanmeldung 10 2011 005 867.2-54 (eingereicht 03/2011) und internationale Patentanmeldung PCT/EP2012/054748
- M. Daschner de Tercero, A. Dresel, U. Fehrenbacher, T. Hirth, U. Teipel
  *Kontinuierliche Herstellung von oxidischen Nanopartikeln in überkritischem Wasser*
  Produktgestaltung in der Partikeltechnologie, Band 4, Fraunhofer Verlag (2008), ISBN 978-3-8167-7627
- M. Daschner de Tercero, J. Schubert, U. Fehrenbacher, M. Türk, U. Teipel
  *Kontinuierliche Herstellung von Eisenoxidnanopartikeln in überkritischem Wasser*
  Produktgestaltung in der Partikeltechnologie, Band 4, Fraunhofer Verlag (2008), ISBN 978-3-8167-7627

# Maren Daschner de Tercero

geboren am 16. Februar 1976
in Hamburg

verheiratet
ein Kind (9 Jahre)

## Bildungshintergrund

| | | |
|---|---|---|
| 05/2008 - 02/2014 | **Doktorandin**<br>Institut für Technische Thermodynamik und Kältetechnik, Fakultät für Chemieingenieurwesen und Verfahrenstechnik<br>Karlsruher Institut für Technologie (KIT) |
| 10/2006 - 11/2007 | **Postgraduales Fernstudium in Nanobiotechnologie**<br>Technische Universität Kaiserslautern |
| 08/2001 - 07/2003 | **Auslandsstudium im Rahmen eines Fulbrightstipendiums: Masterstudium des Chemieingenieurwesens**<br>University of Wisconsin-Madison, USA<br>Abschluss: M.Sc. in Chemical Engineering |
| 10/1998 - 08/2001 | **Studium der Verfahrenstechnik**<br>Technische Universität Hamburg-Harburg<br>Vordiplompreis |
| 04/1997 - 09/1998 | **Grundstudium der Philosophie und Soziologie**<br>Universität Hamburg |
| 08/1982 - 05/1996 | **Grundschule und Gymnasium in Hamburg**<br>Abschluss: Abitur |

## Beruflicher Hintergrund

| | |
|---|---|
| seit 01/2012 | **Projektträger Karlsruhe, Bereich Wassertechnologie, Karlsruhe Institut für Technologie (KIT),**<br>Eggenstein-Leopoldshafen<br>Projektbevollmächtigte |
| 08/2005 - 09/2011 | **Fraunhofer-Institut für Chemische Technologie, Abteilung Umwelt-Engineering**, Pfinztal<br>Wissenschaftliche Mitarbeiterin<br>*Schwerpunkt: Nanopartikelherstellung in überkritischen Fluiden* |

| | |
|---|---|
| 04/2004 - 07/2005 | **Universität Stuttgart, Institut für Technische Thermodynamik und Thermische Verfahrenstechnik**<br>Wissenschaftliche Mitarbeiterin<br>*Schwerpunkt: Biokalorimetrie für die chromatographische Aufreinigung von therapeutischen Antikörpern* |
| 08/2003 - 12/2003 | **BASF Aktiengesellschaft, Abteilung Technische Entwicklung,** Ludwigshafen<br>Praktikantin, *Theoretische Untersuchungen zur Abgasreinigung mittels Absorptionskolonnen* |
| 10/1999 - 05/2001 | **Technische Universität Hamburg-Harburg, Arbeitsbereich Biotechnologie**<br>Studentische Hilfskraft, *Labortätigkeiten* |
| 08/1998 - 09/1998 | **Fehrmann GmbH**, Hamburg<br>Praktikantin, *Arbeiten in der Gießerei* |
| 02/1997 - 03/1997 | **Eppendorf-Nethler-Hinz GmbH**, Hamburg<br>Praktikantin, *Arbeiten in der Lehrwerkstatt und im Betriebslabor* |